# Their Day in the Sun
*Women of the Manhattan Project*

*In the series*

LABOR AND SOCIAL CHANGE

*edited by* Paula Rayman and Carmen Sirianni

# Their Day in the Sun

## Women of the Manhattan Project

RUTH H. HOWES AND
CAROLINE L. HERZENBERG

*Foreword by*
ELLEN C. WEAVER, PH.D.

TEMPLE UNIVERSITY PRESS
Philadelphia

Temple University Press, Philadelphia 19122
Copyright © 1999 by Temple University
All rights reserved
Published 1999

♾ The paper used in this publication meets the requirements of the American National Standard for Information Sciences—Permanence of Paper for Printed Library Materials, ANSI Z39.48-1984

Library of Congress Cataloging-in-Publication Data

Howes, Ruth (Ruth Hege)
   Their day in the sun : women of the Manhattan Project / Ruth H. Howes and Caroline L. Herzenberg ; foreword by Ellen C. Weaver.
     p.   cm. — (Labor and social change)
   Includes bibliographical references and index.
   ISBN 1-56639-719-7 (cloth : alk. paper) — ISBN 1-59213-192-1 (paper : alk. paper)
   1. Manhattan Project (U.S.)—History.  2. Women scientists—United States.  I. Herzenberg, Caroline L., 1932–   . II. Title. III. Series.
QC773.3.U5H68  1999
355.8'25119'0922—dc21                                   99-17683
                                                                                CIP

ISBN 13: 978-1-59213-192-1 (paper : alk. paper)

012909P-2

Printed and bound in Great Britain by Marston Book Services Limited, Oxford

# Contents

| | | |
|---|---|---|
| | Foreword | vii |
| | Prologue | 1 |
| 1 | The Great Scientific Adventure | 6 |
| 2 | The Founding Mothers: Pioneers in Nuclear Science | 20 |
| 3 | The Physicists | 35 |
| 4 | The Chemists | 67 |
| 5 | Mathematicians and Calculators | 93 |
| 6 | Biologists and Medical Scientists | 115 |
| 7 | The Technicians | 132 |
| 8 | Other Women of the Manhattan Project | 152 |
| 9 | After the War | 181 |
| | Epilogue | 201 |
| | Appendix 1: Female Scientific and Technical Workers in the Manhattan Project | 203 |
| | Appendix 2: Chronology | 219 |
| | References | 237 |
| | Index | 253 |

*Photographs follow pages 60, 110, and 174*

ELLEN C. WEAVER, PH.D.

# Foreword

RUTH AND CAROL'S BOOK fills a very personal need for me. I worked on the Manhattan Project in Oak Ridge, but I don't recall any other female scientists working in my groups, and I didn't get to know any of the women working elsewhere at the site.

One reason for this was, of course, that women were a small minority on the project. Another is that for security reasons, the various operations at Oak Ridge were spread out over the site and kept separate as much as possible. I met a couple of women in the locker room, where we changed into protective clothing, but we didn't form lasting friendships. My buddies were all men. Indeed, I'm ashamed to say that I occasionally heard men at Oak Ridge discuss female scientists in insulting terms—concerning appearance, not competence—but that I did not rise to their defense. I remember no sense of camaraderie among women; indeed, we viewed other women with some suspicion. The support that women give one another today did not exist then.

I had majored in chemistry in college, not out of any love of science but because I wanted a skill that would empower me to support my husband, a physicist, through graduate school. When he was spirited off to a secret project in Oak Ridge, Tennessee, I followed as soon as was practicable. After a short stint cranking out calculations on those old Marchant calculators (talk about boring!), I was put to work in a lab with a real project of my own, and just loved what I was doing. Challenges came along daily; it was fun solving them and getting answers. I was only a bit player in the science of the Manhattan Project, but I was a player.

One task I was given was to find a way to shield the radiation of phosphorus-32. This was a beta-emitter, and lead bricks let radiation through. I made "sandwiches" of alternating lead and aluminum layers, but to test the effectiveness of various arrangements I had to assemble a Geiger tube. This was done by splitting mica under benzene with a needle, gluing a thin sheet of mica to the rim of the tube, evacuating the air from the tube, filling the tube with the appropriate gas mixture, and then calibrating it. Subsequent generations never really knew the joy of getting an instrument to work if they hadn't had some experience with the "love and string and sealing wax" that made it function. One of the standard pieces of advice I've given to young scientists—especially women—is to understand the insides of the instruments they rely on and to be able to troubleshoot them.

Later I worked with David Hume's group on devising methods for the quantitative analysis of fission products from the Oak Ridge graphite pile. Again, this was essentially research, and challenging. In addition, a number of the noted scientists on the project taught courses at Oak Ridge. One could get college credit for these from the University of Tennessee, at the bargain price of one dollar per semester hour! I took and passed quantum mechanics and vector analysis, but never tried to transfer the credits toward a degree.

It was an exciting time, and I regret that I met so few of the women who shared it with me (although in later years Mina Rees became a close friend). I'm thrilled to learn through Ruth and Carol's words about so many of the remarkable women who contributed to that great enterprise. This book lets us meet one another at last and to swap stories. The authors have done superb detective work, tracking down an impressive number of us—more than 300.

When I read the personal reminiscences of the men who were nuclear pioneers, I'm struck by the importance they place on their friends and associates, and on the often intense interaction among them, which could lead to major insights in both the theory and practice of science. And I'm a bit jealous. By and large, women did not share in that give-and-take. Through this volume, I can vicariously share in it, and at least indulge in some "might-have-beens."

I hope that *Their Day in the Sun* will find many readers, women and men, girls and boys. Women were players in the Manhattan Project—essential players. They deserve honor and remembrance.

*Ellen C. Weaver, Ph.D., is Past President, Association for Women in Science.*

# Prologue

ON JULY 16, 1945, the light of the first nuclear explosion stabbed through the New Mexico predawn darkness. Three weeks later, an atomic bomb devastated the Japanese city of Hiroshima, and the American public learned for the first time of the vast research effort codenamed the Manhattan Project. The project's scientists and engineers suddenly found themselves front-page news.

Contemporary stories in the press about the bomb and the people who had created it featured only one woman, Lise Meitner, a noted Austrian physicist who had contributed significantly to the discovery of fission just before World War II but who did not actually work on the Manhattan Project. She was usually relegated to a human-interest sidebar.

Photographs showed male project leaders wearing business suits and military uniforms, male scientists in white lab coats, and GIs sweating in T-shirts in the New Mexico desert or on the island of Tinian in the South Pacific. An occasional woman appeared as a secretary in a group seated around a table or as a WAC (a member of the Women's Army Corps) operating a switchboard. The earliest books that came out on the Manhattan Project, including official histories, made no mention of contributions by female scientists and engineers.

Was there something wrong with this picture?

Like many other American women, the authors have relatives who fought in World War II or were otherwise involved in the war effort. As children, we were deeply affected by stories of the war and fascinated by descriptions of the atomic bomb. What we read and heard influenced our eventual decisions on educations and careers. But we still knew very little about this immense project.

We both became physicists and have been privy to professional gossip, but we accepted the story that the development of the atomic bomb had been an exclusively male activity. There seemed no reason to doubt it. Although we had known about Leona Marshall, a physicist who as a new Ph.D. had contributed to the creation of the first controlled, self-sustaining, nuclear chain reaction, we had little idea that other women had made substantial technical contributions.

This book began almost by accident. We needed to provide a chapter for a volume on women's role in the use of military force. Although we were aware that wartime personnel shortages had opened opportunities in scientific and technical fields to women, we expected to find very few women who had been involved in technical aspects of the Manhattan Project, and colleagues' reactions worked to confirm our expectations: "It shouldn't take very long to complete the study, since there weren't any women"; "What women? There weren't any"; and "It's a null set." If we had listened to the advice of these otherwise knowledgeable colleagues, we might have written a very short chapter and considered our responsibilities fulfilled. However, we had heard rumors that some women did indeed work on the Manhattan Project, and we were intrigued by the challenge of finding them—if they existed.

As scientists, we had experience in tackling unanswered questions and in applying science to resolve problems in the world around us. This, however, was a new type of challenge—more like a detective story than a scientific investigation. It was something for which our professional training had not prepared us. We both like puzzles, however, and we set eagerly to work. We visited the sites of the Manhattan Project. We searched the relevant records. We interviewed witnesses and pieced together all this information. We had thought ourselves well informed on the topic of female scientists, but our early findings brought a real surprise. Despite their absence from the early reports, women worked on every aspect of the Manhattan Project, except combat operations. Wartime personnel shortages opened scientific and technical opportunities that were not available to women in peacetime, and women took them.

We began our study in a very casual manner at a meeting of the American Physical Society by simply asking our friend Melba Phillips, a distinguished physicist and historian of physics, whether she had worked on the project. We knew that she had done graduate work with J. Robert Oppenheimer, the physicist who had led the effort to design and build the atomic bomb at Los Alamos. Phillips replied that she had not worked on the bomb (characteristically neglecting to mention her leadership in opposing the nuclear-arms race), but she suggested that we speak to Naomi Livesay French, who was also at the meeting. French, who had worked at Los Alamos calculating mathematical models for bomb design, spent four hours the next morning cheerfully answering Ruth's questions about her experiences. She also provided us with the names and addresses of women who had worked with her. These women responded willingly to telephone calls, and we quickly developed an initial network of women who had worked on the science and technology of the Manhattan Project.

We collected most of the material for this book by following the links of a number of old-girl networks such as this one, largely through telephone interviews. In this manner we reached women associated with all the major Manhattan Project sites, including Oak Ridge, Hanford, and Los Alamos. In addition, we sent a questionnaire to all the participants in the Los Alamos Fortieth Reunion whose names were obviously female. Funding for the questionnaire and some of the telephone calls came from a research grant from Ball State University. We searched the published literature and materials accessible on the Internet. As we published preliminary accounts of our work and spoke about it at meetings, members of our audiences began to contact us with more information, which we have followed up as well as we could.

We have information on more than 300 women—half of whom are now dead—who performed technical work on the Manhattan Project, but we are certain that those 300 women are only a fraction of the women associated with the project. Most of the surviving women who contributed are now in their eighties, but many of them retain their vigor, their enthusiasm, and their interest in the world. For example, when we contacted her, Marge DeGooyer was recovering from a broken arm she had sustained while barn dancing. Joan Hinton, who now makes her home in China, had been driving alone across the United States, doing some bird-watching. When her car broke down, she repaired it herself as a matter of course.

Meeting and talking with these remarkable women has been truly exciting. In addition, we have talked with many participants' husbands, children, and friends, whose pride in the work of their relatives and acquaintances has been heartwarming to us as female scientists. Harold Argo proudly told us about his wife's work as a physicist with Edward Teller. Herbert Pomerance wrote to us about his wife's role in the Manhattan Project and responded willingly to our many questions. Will Happer of the U.S. Department of Energy, now a professor at Princeton University, wrote us a long, informative letter about his mother's work as a physician at Oak Ridge during World War II. Many other relatives of the women who did technical work supplied helpful information as well.

We are also very grateful to our many male colleagues who participated in the Manhattan Project or knew those who did, and who patiently responded to our interest in the women who worked with them. Busy men such as Edward Teller and the Nobel laureate Glenn Seaborg made time to talk to us. Some, such as Roy Ringo, Mel Freedman, and Heinz Barschall, submitted to repeated interviews and were tireless with their help. Carol, who lives in Chicago and works at Argonne National Laboratory, was able to talk to a number of individuals with connections to the Metallurgical Laboratory, the precursor of Argonne. Most helpful of all were Jane and Bob Howes, Ruth's mother-in-law and father-in-law, who patiently answered questions about early life at Los Alamos and introduced us to their friends and fellow participants in the project.

For the most part, the few histories that discuss women's contributions to the atomic bomb emphasize their roles as wives, mothers, secretaries, or hostesses. These women's contributions to the Manhattan Project were, indeed, critical to its success. Without the women who handled the flood of paper that mobilized supplies for the laboratories and production facilities, the Manhattan Project would surely have ground to an early stop. The members of the Women's Army Corps who drove the trucks and operated the telephones kept the machinery of the project moving. Female production workers produced necessary fissile and other materials. The contributions of these women should be honored, but they have been recorded elsewhere. (See, for example, the books by Wilson and Serber [1988], and Roensch [1993].) Here we focus on women's work in the laboratories and test sites of the Manhattan Project. These women, unlike their nontechnical colleagues, have been largely ignored.

We do not claim to have discovered all the women who made important scientific and technical contributions to the Manhattan Project. Indeed, we have received many letters and e-mail messages describing the work of women whose names were not yet on our lists. It has been a delightful experience each time we have learned of yet another woman who worked on the project, but it has also kept us aware that there are many others whom we have been unable to locate. What's more, the events described here took place more than fifty years ago; even after a much shorter span of time, different people remember events differently and have varied perspectives on them. These factors remained firmly in our minds as we tried to uncover the events and untangle the strands of a half-century-old secret project, in an effort to bring to light scientific and cultural information that has been largely lost to history.

Even recognizing that our data are incomplete, we believe that what we have discovered makes an exciting story. We have thoroughly enjoyed our detective work, and we hope that some of our readers will go on to examine additional aspects of this intriguing topic.

Finally, we want to note that the women who worked on the atomic bomb had differing reactions to their creation. When the work began, most of them naturally focused on the urgent need to produce a weapon that would end the war. After Hiroshima, they saw their work on the Manhattan Project as a major contribution to the victory, and they celebrated the fact that their husbands, fiancés, and brothers would not have to invade Japan. They are still proud of their contributions to the war effort. Many were nonetheless shocked by the devastation of Hiroshima and Nagasaki, and went on to play active roles in the struggle for civilian control of nuclear power and the disarmament movement that followed the war.

In other words, their reactions, like their contributions to the Manhattan Project, were as diverse as those of the men who worked beside them.

# 1   The Great Scientific Adventure

THE MANHATTAN PROJECT developed the field of nuclear physics and produced a weapon whose power, fifty years later, continues to shape the politics of the world. It spawned entire industries and new areas of science, and the style of industrial-scale research developed during the Manhattan Project forecast the nature of scientific enterprise in the United States.

The popular heroes of the project were a group of theoretical physicists, all male, whose arcane knowledge of quantum mechanics and the structure of the atomic nucleus first hinted at the possibility of a weapon based on nuclear fission. In fact, these men worked far from alone. Without experimental scientists and engineers located throughout the country and a massive industrial effort overseen by the military, the atomic bomb would never have been created. The people who conducted this research effort reflected the makeup of the nation. They came from all races and religions, and many of them were women. In fact, it was a woman, Lise Meitner, who provided the first theoretical explanation of nuclear fission—and coined the term.

The important scientific and technical work that women contributed to this huge national undertaking must be understood in context. Thus we begin their story with an overview of the secret, high-pressure American effort to develop an atomic bomb: the Manhattan Project.

## GENESIS OF THE MANHATTAN PROJECT

In Sweden, during the 1938 Christmas vacation, two German physicists, Lise Meitner and Otto Frisch (Meitner's nephew), examined data from Otto Hahn and Fritz Strassmann's Berlin laboratory, where Meitner had been an active participant until Hitler's government had forced Jewish scientists into exile. Now she interpreted the surprising results of bombarding uranium samples with neutrons. She realized that the barium found in the irradiated samples could be produced only if the neutrons split uranium nuclei into two pieces. Meitner called the splitting of the nucleus "fission," becoming the first to use the word in a nuclear context. Today it has hardly any other context.

At the end of Christmas vacation, Frisch reported the discovery of nuclear fission to Niels Bohr, the Danish Nobel laureate. Bohr carried the news across the Atlantic to the Fifth Washington Conference on Theoretical Physics, held in January 1939. The physicists immediately realized, as Meitner had, that the energy released by each nuclear fission reaction was millions of times greater than the energy released in the atom-atom reactions of chemistry. If the fission of a uranium nucleus produced neutrons, as well as large fragments that were the nuclei of lighter elements, these secondary neutrons could be used to trigger other fissions. Nuclear fission could then form a self-sustaining chain reaction, releasing immense amounts of energy. The scientists set to work to study it in detail. Within a year, the international physics community published more than 100 papers on nuclear fission (Rhodes 1992).

The possibility of using this vast source of energy as a weapon attracted the interest of military developers around the world. By the end of April 1939, Germany had begun a research program aimed at developing an atomic weapon. The British project started more slowly, but was in high gear by 1940. Work was under way in the Soviet Union before being curtailed by the Nazi invasion of 1941 (Rhodes 1992). In Paris, Irène Joliot-Curie, daughter of Marie Curie, and Irène's husband, Frédéric Joliot, both Nobel laureates, headed up France's program.

These early projects, though aimed at building a nuclear explosive, were primarily concerned with understanding the basic physics of nuclear fission. But even after it became clear that nuclear fission might have military significance, the results of these projects continued to be published in the open literature. Scientists in the United States and Britain instituted a voluntary ban on publication, but the French group refused to cooperate because their results had so little direct application to a weapon.

In the United States, physicists determined that a fission chain reaction was indeed possible, but they faced a number of difficulties in producing one. Only one isotope of uranium underwent fission, and that isotope, uranium-235, accounted for only about 0.7 percent of naturally occurring uranium. Separating the amount of the chemically identical isotopes of uranium needed for a reaction would prove a challenge in the laboratory and would require a near-miracle on an industrial scale.

In March 1939, the Italian Nobel laureate Enrico Fermi, who had emigrated to the United States with his wife, who was of Jewish extraction, presented his results to a group of Navy technical experts. Fermi's calculations suggested that a chain reaction could release an enormous amount of energy, but he had no experimental evidence that a chain reaction was even possible, and the problems involved in producing enough uranium-235 for a weapon were daunting even on paper. The Navy did not consider nuclear fission promising, although its experts asked to be kept informed of experimental progress (Rhodes 1986).

The Hungarian physicist Leo Szilard, also a refugee from Hitler's conquest of Europe, recognized both the military potential of nuclear fission and the urgency of the Nazi threat. With his fellow Hungarian emigrants Edward Teller and Eugene Wigner, he persuaded Albert Einstein to sign a letter to President Roosevelt, alerting the President that a superweapon was possible and that the Germans might be working on one. Alexander Sachs, an informal adviser to Roosevelt, brought him the letter in the fall of 1939 and convinced him that the United States needed to explore the possibility of a fission weapon.

Roosevelt's response was to appoint a Uranium Advisory Committee. The first funding for the project, $6,000 for the purchase of uranium and graphite, reached Columbia University in February 1940. In June of that year, the Uranium Committee became a subcommittee of the new National Defense Research Committee, which was headed by Vannevar Bush. Intelligence sources had learned that the Germans had begun their own studies of nuclear fission.

In November 1940, a $40,000 contract from the National Defense Research Council went to Columbia University to support the demonstration of a self-sustained chain reaction and to make further measurements on nuclear parameters. A year later, sixteen contracts for about $300,000 had been awarded to various schools, among them Columbia, Princeton, Cornell, Johns Hopkins, Chicago, California, Minnesota, Virginia, and Iowa State College (later to become Iowa State University), and to such organizations as the Carnegie Institution, the Standard Oil Development Company, and the National Bureau of Standards. The projects varied widely: studies of the basic physics of nuclear fission, experiments on ways to separate uranium isotopes, purely theoretical work, and more. The Manhattan Project was under way.

## The Expanding Project

The uranium effort was minute compared with the radar project at MIT, which had a budget of several million dollars in the fall of 1941 (Smyth 1977). But as it became clear that an atomic bomb actually could be built, the Manhattan Project absorbed more money and more people. As the project matured and its focus narrowed, the smaller research groups were absorbed by the bigger centers. Designers required larger facilities for production; sites had to be isolated to protect the population in case of accidents; and project leaders, increasingly concerned with security, wanted a smaller number of closed project sites where they could protect classified information.

On September 23, 1942, General Leslie Groves, who had just finished directing the construction of the Pentagon, assumed command of the Manhattan Project. Groves had hoped for an overseas assignment, but he accepted command of the Manhattan Engineer District—the project's official name—because he was convinced that its success would be critical to winning the war.

General Groves managed the project as a military mission, and his abrupt style infuriated many of the scientists. But without his willingness to make risky decisions about technical issues and personnel, the atomic bomb would never have been finished before the end of the war. He insisted, for instance, that all three possible methods of separating the uranium isotopes move into production while the Hanford reactors were being constructed to produce plutonium—and before it was known whether any of the three methods would work!—and he appointed Robert Oppenheimer to head the bomb design effort in spite of the physicist's left-wing politics and his lack of management experience. Both decisions were inspired. The general transformed the Manhattan Project from a research effort into an industrial enterprise.

At its height, the Manhattan Engineer District consisted of three major sites. Clinton Engineer Works, at Oak Ridge in the Tennessee mountains, was the site of major industrial facilities to separate uranium isotopes. Hanford, near Richland, Washington, held the huge nuclear reactors designed to produce the new fissile element plutonium and the enormous remotely operated facilities to separate the plutonium from the uranium reactor fuel. In Los Alamos, on a remote New Mexico mesa, physicists and engineers assembled to work on the actual design and construction of the bomb.

Major subprojects at the University of California at Berkeley and the Metallurgical Laboratory of the University of Chicago developed methods for separating uranium isotopes and producing plutonium metal on a laboratory scale. These were then implemented on an industrial scale at Clinton and Hanford. Albert Nier's group at the University of Minnesota pioneered the separation of uranium isotopes using highly developed mass spectrometers. Having produced the first samples of enriched uranium-235 for critical nuclear experiments, Nier and his group moved to Kellex, a subsidiary of the M. W. Kellogg Company established to oversee isotope separation.

The Ames Project at Iowa State College, directed by Frank Spedding, supplied many of the chemically pure materials vital to the success of other phases of the Manhattan Project. It produced, among other things, the uranium metal "eggs" used in reactor CP-1: Chicago Pile Number One, the reactor built by Enrico Fermi's group under the University of Chicago's football stands, where the world's first controlled, self-sustaining nuclear chain reaction took place. Throughout the war, the Iowa State site served as a production facility for very pure uranium and other materials needed to make the bomb.

John Dunning's group at Columbia University attacked the problems associated with gaseous diffusion and studied other possible methods of isotope separation. The results guided the design of the production facilities at Clinton. Among other contributions, the Columbia group and the industries charged with building the production plant solved the pivotally important problem of producing screens that would not corrode or clog during the passage of highly reactive uranium hexafluoride gas.

In addition, other companies and schools undertook pieces of the project work, and female scientists and technicians could be found at most of these locations.

## Finding the Manpower—and Womanpower

In less than four years, during a major war, the Manhattan Project evolved from an eclectic research effort based on untested ideas in theoretical physics into a major industrial enterprise with facilities spread across the North American continent, consuming a significant portion of U.S. spending. Just before the United States entered World War II in December 1941, the Manhattan Engineer District consisted of 16 projects with a total budget of $300,000 (Groves 1962). Counting the graduate students and technicians assisting the principal investigator on each project, the entire staff may have been as large as 200 people. By the time the first atomic bomb was dropped on Hiroshima, the Manhattan Project had expended $2.2 billion on facilities and production and employed up to 130,000 people at a time (Gosling 1990).

Wartime labor shortages were universal. By 1943, the Army was 200,000 men short of the number needed for the D-Day invasion, and the average age of the GIs had climbed above twenty-five. Defense projects needed 315 more physicists than were available (Kevles 1978). The managers of the Manhattan Engineer District seem to have recognized that all technical personnel, regardless of sex, should be considered for recruitment to the project.

University labs provided the first source of scientists and technical personnel. Scientists and engineers who had been studying nuclear fission since 1939 were immediately recruited, and so was their equipment. Los Alamos, needing information on the nuclear properties of uranium, borrowed accelerators from universities and took the operators along with the machines. The University of Wisconsin's Van de Graaff accelerators disappeared almost overnight. The physicists who worked on them followed, with minimal explanation to friends and relatives.

Joan Hinton, a graduate student in physics at Wisconsin, had been wondering why faculty and graduate students had vanished. Then she was offered work on a war project in New Mexico. After accepting, she went to the university library and checked out a book on the state. The names of all her missing colleagues were on the borrowers' card in the back of the book, so she was not surprised to find friends when she arrived in Los Alamos (J. Hinton 1990).

The Manhattan Project competed for resources with research on radar, underwater sound (needed to protect convoys from German submarines), proximity fuses, and many other wartime projects that pulled technically trained personnel from the university labs. When recruiters found a scientist with skills they needed, sex didn't matter. Many women with technical training who had not worked outside the home taught university classes when male faculty and graduate students left for war projects. (Industrial scientists whose work was considered essential to the war effort generally received deferments and remained at their jobs.)

Once people had been hired for war-related jobs, they needed permission to leave or change employment. With such a drain on technical labor, women found themselves being recruited by organizations that would have scorned their applications in peacetime. In spite of her German background, Maria Mayer, for example, received a part-time position working on the uranium project at Columbia University instead of serving as an unpaid "volunteer" professor.

Refugees provided a rich source of expertise. American universities snapped up the Nobel laureates Albert Einstein and Enrico Fermi. Less famous scientists such as Leo Szilard, Edward Teller, and Eugene Wigner found important positions as well. Fermi led the experiments on nuclear fission that eventually produced the first sustained chain reaction. Nathalie Goldowski, who had served as chief of metallurgy for the French Air Force before fleeing France, worked in Chicago on preventing the corrosion of the aluminum jackets that surrounded the uranium fuel in plutonium production reactors.

Not all refugees found an eager welcome on secret wartime projects, however. Military security rejected some whose families remained in occupied territory and a few who were considered security risks in their own right. Such outstanding young scientists as Stanislaw Mrozowski, a Polish molecular spectroscopist, and his wife, Irena, came to study at the University of Chicago before Poland was occupied. With close family trapped behind Nazi lines, they were not allowed to do secret work. They remained at the University of Chicago as instructors during the war (Mrozowski 1991). The Manhattan Project badly needed Elizabeth Rona's expertise on polonium, but because her family remained in Hungary, she had to help the project unofficially—often via a friend who would ask, "I was wondering whether you knew...." Her formal appointment didn't come through until 1947. Clearance procedures were inconsistent, often depending on who knew the scientist in question and how valuable his or her expertise seemed. Some scientists who worked on the Manhattan Project, such as Goldowski, were later unable to obtain security clearances in the environment of the cold war.

In the early days, British research on nuclear explosives outstripped work in the United States. Scientists communicated across national lines, and formal agreements eventually authorized exchanges of personnel and information. The British delegation at Los Alamos performed valuable technical work on the first atomic bomb. (Unfortunately, the delegation also included the physicist Klaus Fuchs, who passed important details of the research to the Soviet Union.)

But the Manhattan Project needed more than scientists and engineers. Machinists, pipe fitters, and other skilled laborers were even rarer than physicists in an economy gearing up for war. General Groves writes of the cooperation of labor unions in his effort to secure personnel for his projects. Du Pont moved the labor forces of entire plants from Oklahoma to Hanford to supply skilled workers to build the production reactors and separation facilities. The hungry project finally persuaded the Army to establish the Special Engineer Detachment (SED) so that able technical personnel who had been drafted could be assigned to research work. A detachment of the Women's Army Corps was deployed to each of the major laboratory sites. Uniformed personnel, including WACs, played key roles on the scientific and technical staff of the laboratories.

The military personnel were not always thrilled with these arrangements. Many of them, including many WACs, had volunteered for overseas duty and were understandably annoyed to find themselves babysitting a bunch of long-haired scientists. Furthermore, military pay scales were well below civilian for the same scientific jobs, and civilians didn't have to worry about inspections or drill.

According to Edward Wilder Jr., a Navy lieutenant sent first to Clinton and then to Los Alamos to design and manufacture the explosive lenses for the implosion bomb, the military workers believed that they had been assigned the explosives work because it was considered too dangerous for civilians (Truslow 1991, Appendix). Military personnel, unlike civilians, were not allowed to bring their families to the project. And by no means the least of the military gripes was that for security reasons, soldiers stationed at Los Alamos were forbidden to apply to Officer's Candidate School (Fitch 1974).

On the basis of numbers alone, women were important at all the project sites. They accounted for 9 percent of the 51,000 employees at Hanford in 1944, when the site's staff was at its largest (Williams 1993). At Los Alamos, where tight security made it difficult to hire people who did not already live on the site, opportunities for women may have been even greater. In September 1943, some sixty women worked in the Technical Area at Los Alamos. By October 1944, about 30 percent, or 200 members of the labor force in the Tech Area, the hospital, and the schools were women (Hoddeson et al. 1993). Of these, twenty could be described as scientists and fifty as technicians. Fifteen women worked as nurses, twenty-five as teachers, and seventy as secretaries or clerks.

Although many women's precise job titles at Los Alamos remain unknown, rough numbers show about twenty-five of them working on chemistry and metallurgy, twenty on bomb engineering, sixteen on theoretical physics, four on experimental physics, eight on ordnance, and four on explosives. Two women worked with Enrico Fermi, who had moved to Los Alamos when it opened in 1943. These numbers are given by divisional assignment instead of by job title (Henriksen 1994), so a few of these women may have held clerical jobs, but it's clear that most of them were scientists or technicians.

The number of women working on the Manhattan Project contrasts sharply with the Apollo Project of the 1960s, which was comparable in size and scope. At its peak in 1965, when Apollo engaged 5.4 percent of the national supply of scientists and engineers (Levine 1987, Table 4), women accounted for only 3 percent of NASA's scientific and engineering staff (Fries 1992, Appendix B, Table 8).

## Life Inside the Manhattan Project

Once families had moved to the remote, secret facilities of the Manhattan Project, wives who could were strongly encouraged to join the work. Groves writes of Los Alamos:

> The system was designed to encourage the wives of our people to work on the project, for those who worked obtained priority on household assistance. Some of the wives were scientists in their own right, and they, of course, were in great demand, but with labor at a premium we could put to good use everyone we could get, whether as secretaries or as technical assistants or as teachers in the public school that we started for the children.
>
> To enable the mothers of young children to work, a nursery school was organized on a partially self-supporting basis; its financial losses were carried by the government. The elementary and high schools were operated as free public schools, with all expenses borne by the project. (Groves 1962, p. 166)

"In the early days of the Project, most of the women who worked in the [Tech] Area were wives of young physicists and chemists," wrote Charlotte Serber, technical librarian at Los Alamos and herself the wife of a physicist. "In a great many cases, they had not intended to work when they came to Los Alamos. But the force of social pressure and the obvious need for all hands, trained or untrained, brought most of them rapidly onto the payroll" (Serber 1988, p. 57).

Not all wives eagerly embraced the opportunity, however. Peggy Hemmindinger held a bachelor's degree in physics and had actually done weapons-related work, but when she followed her husband to Los Alamos she did not work outside her home (Manley 1990). Eleanor Jette, who was trained in mathematics, recalls her indignant response to an attempt to recruit her to work in the Tech Area at Los Alamos, and to the recruiters' comments about her patriotism when she refused: "I discussed the events of the autumn and their effect on my well-being. I expressed my opinion of living conditions in Los Alamos, particularly the fire hazards. I was sure the wives, who should be at home taking care of the children, were intimidated or encouraged to work because the penny-pinching Washington administration didn't want to build housing. I theorized that properly qualified teachers weren't hired for the same reason" (Jette 1977, p. 29). Even in wartime, it was accepted that the duties of motherhood superseded the national need.

Working wives frequently expressed concern about the effect on their families, who were, of course, considered the woman's responsibility. "For the potential working wife, there was one chief worry," wrote Charlotte Serber. "Could she manage her home here on the mesa and work too? Would her home life suffer? Would her husband be neglected? Would her children become delinquents? Would it be any more difficult than working a forty-eight-hour week in a city?" (Serber 1988, p. 67)

Despite their domestic concerns, many wives did work, motivated by patriotism, curiosity about the work that engrossed their husbands, and perhaps a touch of boredom in the isolated locations of the project. Women were not promoted or paid in proportion to their efforts, however. "For the working wife, the actual process of being hired in was not very complicated," Serber writes. "It entailed filling in a multitude of forms, getting a pass to the Tech Area, listening to a speech on security, hearing an oversimplified version of working conditions on the Hill, and getting her salary set. Working conditions included forty-eight hours to the work week, two weeks of vacation with pay, sick leave, one day off a month for a shopping trip to Santa Fe, maids available through the Housing Office, and a nursery school for her children. The working wife's salary, which was set very arbitrarily, was influenced less by her previous work history than by the fact that she really had no bargaining power. She lived, after all, in a sort of company town" (Serber 1988, p. 67).

Life at Los Alamos, Clinton, or Hanford proved rigorous for most recruits. Chronic housing shortages, limited services, and the remoteness of the locations particularly bothered many of the Europeans, who had been accustomed to servants and an urban setting. Unpaved streets sent clouds of dust into windows. Water shortages and power outages plagued communities that rapidly outgrew their original designs.

The standard workweek was forty-eight hours, but many people put in considerably more time. The project dominated ordinary life. "Los Alamos was like a giant ant hill," writes Eleanor Jette. "The atom bomb was its queen and the Tech Area was her nest" (Jette 1977, p. 42).

Tight security shut residents who were not working on the project out of the mainstream of the community's existence. When crises broke in the technical work, the entire community suffered. Husbands remained in the labs, and wives who did not work in the Tech Area received no explanations for missed dinners and appointments. They were shut completely out of the work that consumed their husbands' lives.

Still, crises were part of the excitement of the work. "The laboratories had a cluttered, disorderly, academic air," Serber writes. "The offices were simple enough, though incredibly dirty, overcrowded, and badly equipped. Physically, the Tech Area was certainly not a very unusual place. But it did have a spirit which was strange. Its tempo was too fast; its excitement was almost too high. The Area was in a state of continuous crisis, and it soon became clear that speedup was its permanent tempo and excitement its permanent mood. This hyperthyroid quality was contagious and soon, in each newcomer to the Area, any disappointment with its physical drabness was rapidly followed by a real enthusiasm for both its work and its personnel" (Serber 1988, p. 57).

Security dominated other considerations. At Los Alamos, scientists and their families were told to report to unknown destinations and could not explain their work to relatives. Their address was a post-office box. If a family member left the project, even to attend college or for some other legitimate reason, he or she could not return to the closed city. Mail was censored. Though only Los Alamos was actually an Army camp, the Army ran Hanford and Clinton as well. The comforts and complaints of personnel played major roles only when ignoring them threatened the supply of skilled labor needed for the technical work on the bomb.

Hanford and Clinton were not military bases, and both workers and relatives could come and go. General Groves became so concerned about the turnover of clerical staff at Hanford that he hired Buena Maris, the Dean of Women at Oregon State College, to serve as Supervisor of Women's Activities. "Admittedly, our concern with morale was not entirely altruistic, for a stable clerical force was essential," Groves wrote. "We simply could not afford a constant turnover. The trouble was that employees found it easy to get jobs in Yakima, Seattle and other nearby cities where living conditions were far pleasanter. The turnover hazard started on arrival" (Groves 1962, p. 90).

Buena Maris and Margaret M. Shaw were the only two women in top management at Hanford during the war. Maris had full access to the management of Du Pont and the Army commanders at Hanford. She was able to deliver such minor miracles as asphalt sidewalks to save rationed shoes and a late bus home from Pasco so that shoppers would not have to contend with drunken construction laborers. According to her daughter, Marjorie Peterson, Maris worked eighty-hour weeks for a salary of $4,000 per year and lost a great deal of weight doing it. "She'd come home exhausted and say: 'What I need is a good wife,'" said Peterson (Williams 1993, p. C8).

Despite the working conditions, women were attracted to the work of the Manhattan Project for a number of reasons—foremost among them, of course, the desire to contribute to the war effort. Their husbands and brothers were in combat overseas, and the women wished to make whatever contributions they could. The rise of Nazi Germany and the growing documentation of genocide in Europe convinced most Americans that winning the war should outweigh any reluctance to work on weapons.

The attitude that glorified Rosie the Riveter also infused the scientific community. Women were urged to free men up to fight by taking on work they otherwise would not have done. Other women simply needed jobs to support themselves. Some chose technical work because it offered a chance to acquire new skills. Marge DeGooyer, who had been trained as a secretary, was working as a taxi driver before joining her parents at Hanford. She applied for a clerical job at the lab, where supervisors noticed her high math skills. They told her that a job in the technical area would teach her more than getting a college degree in chemistry. She took the technical job and later became a technician at Hanford in charge of setting up and maintaining mass spectrometers (DeGooyer 1995). The wartime labor shortages opened opportunities to women that had not existed before the war and would not exist once the war was over.

Wartime urgency did not remove glass ceilings. Few women joined the top management of the project, and there is some evidence that they were excluded on the basis of sex. Certainly the military would have found it inconceivable to deal with a female Oppenheimer. Peter Yankovitch, a graduate student in chemistry at the University of California at Berkeley, recalls taking a job overseeing the work of a shift in the analytical chemistry labs. He and two other young male graduate students supervised women with far more experience, who actually taught them the techniques involved (Yankovitch 1995).

Many able women occupied unpublicized positions just below the top echelon of the National Defense Research Committee and its wartime successor, the Office of Scientific Research and Development (OSRD). Mina Rees, who had a Ph.D. in mathematics, served as technical aide and executive assistant to the chief of the applied mathematics panel in both organizations from 1943 to 1946, when she became head of the mathematics board of the Office of Naval Research (*American Men and Women of Science*).

Gladys Amelia Anslow studied nuclear physics at Yale while serving as an assistant professor of physics at Smith College. She was the first woman to work on the 8 MeV cyclotron at Berkeley. From July 1944 to December 1945, she served as chief of the Communications and Information Section of the Office of Field Service of the OSRD, controlling the flow of classified information to the research community. For her service, she received the Presidential Certificate of Merit (*Smith Alumnae Quarterly* 1945; Fleck 1993).

Female CEOs in industry were as scarce as their counterparts in the science-policy arena. One notable exception was Mrs. H. K. Ferguson, who headed Ferguson Engineering, a company founded by her husband. The Cleveland company was hired in June 1944 to construct the huge thermal-diffusion plant at Oak Ridge, Tennessee, following a model plant developed at the Philadelphia Navy Yard. The gaseous-diffusion plant seemed to be in serious difficulty, and Groves had decided to construct the thermal-diffusion plant as an alternative.

The installation was to consist of 2,142 double-walled tubes, each forty-eight feet tall. The tubes were heated to 280 degrees C on the inside and water-cooled externally so that uranium hexafluoride gas forced between the columns would rise on the inside and sink on the outside. The lighter uranium-235 would concentrate slightly in the rising gas, since molecules containing it moved a bit faster than those containing uranium-238. Manufacturing the columns posed such challenges that twenty-one firms turned down the contract. The facility had to be constructed in ninety days. Ferguson, which formed a dedicated subsidiary to accomplish the work, brought the first columns on line in only sixty-nine days (Groueff 1967).

The final phase of the Manhattan Project, the wartime delivery of a completed atomic bomb, known as Project Alberta, was a combat operation. During World War II, the United States barred women from combat, so women played little role in Project Alberta. The research groups for this phase of the project were sent to Tinian, the Pacific island where the bomb was assembled, but their female members remained in the United States. Frances Dunne, who worked with the Explosives Assembly Group at Los Alamos, helped her "boys"—the thirteen tech sergeants with whom she worked—buy civilian clothes for their trip to Tinian, but she stayed at Los Alamos (Dunne 1991).

Project Alberta, however, was the exception. Women participated in and made important contributions to all other aspects of the Manhattan Project, and their stories mirror the story of the project itself.

## 2  The Founding Mothers
*Pioneers in Nuclear Science*

MANY SCIENTIFIC LEADERS of the Manhattan Project emigrated to the United States to escape Hitler's conquest of Europe. Enrico Fermi played a leading role in developing the first self-sustaining nuclear fission reaction. Niels Bohr, the Danish Nobel laureate, worked at Los Alamos, where Hans Bethe, a German physicist, headed the Theory Group. The Ukrainian chemist George Kistiakowsky ran the Explosives Division at Los Alamos. There are many other examples. Without the scientific insight of European scientists, the U.S. research effort would have progressed far less rapidly.

These men arrived in the United States with significant scientific reputations based on their work in Europe. Because of their prominence, the political and military leaders of the Manhattan Project deferred to their expertise, and they are recognized as leaders in the development of the atomic bomb.

Though women are never listed among the leaders of the Manhattan Project, a surprising number of the important discoveries in nuclear physics were made by European women. At the outbreak of World War II, Lise Meitner was considered one of the leading nuclear physicists in Europe, and Irène Joliot-Curie had won a Nobel Prize for the discovery of artificially induced radioactivity. Joliot-Curie's Paris laboratory and Meitner's Berlin research team raced the Italian group led by Enrico Fermi and Ernest Rutherford's laboratory at Cambridge, England, in the search for new principles of nuclear physics and even new chemical elements. Perhaps because the study of the nucleus was still a new, peripheral area of science, women were more welcome than in more established disciplines. But why did women's prominence in nuclear physics not carry over directly to the Manhattan Project? Why did the women who led the development of nuclear science in Europe not join their male counterparts as leaders of the effort to develop the atomic bomb?

## The Women and Their Research

Nuclear science was born in Europe, and the bulk of its early development took place there. The guiding physical ideas arose from university laboratory experiments and theoretical studies. Centers grew up in Göttingen (Germany), Berlin, Paris, and Cambridge. The international nuclear scientific community could fit in a single room, and the scientists exchanged letters describing their work.

Although at least twenty-five women, primarily from Europe or trained there, participated in the early research in nuclear science, there were relatively few female American and Canadian scientists in the forty years before World War II, and even fewer who were involved with nuclear studies—in large part, no doubt, because America and Canada did little work on nuclear science before the outbreak of the war.

Female scientists discovered a number of new elements, some radioactive and some with stable isotopes. Marie Curie and her colleagues first identified both radium and polonium. Ida Tacke, a German chemist, detected the element rhenium when she was only seventeen years old, while working with her future husband, Walter Noddack, and O. Berg (Taton 1964). Ida Tacke Noddack has also been credited with the identification of element 43, technetium (Dampier 1949). Lise Meitner played a role in the discovery of protactinium, and Marguerite Perey, a nuclear chemist and physicist working at the Curie Institute in Paris, was the discoverer of the element francium (Taton 1964).

Many of the same women pioneered the study of the radioactive decay of various elements. Marie Curie was among the first to study the decay of thorium, and she and her husband, Pierre Curie, first examined the radioactive decay of radium (Taton 1964). Harriet Brooks worked with Ernest Rutherford to determine the density of the radioactive gas radon and to study radon decay products. Lise Meitner, working with the chemist Otto Hahn, discovered the first example of nuclear isomeric states in an isotope of platinum (Hampel 1968; Weeks 1956). Irène Joliot-Curie and Catherine Chamié first measured the half-life of radon (Parkinson 1985).

In addition to their discoveries of new elements and new stable and radioactive nuclei, female nuclear scientists discovered many of the characteristics of the new particles involved in radioactive decay. The Curies showed that the charges transported by beta radiation are negative. With the German chemist Fredrich O. Giesel and the French physicist Henri Becquerel, Marie Curie showed that beta radiation consists of high-speed electrons (Taton 1964; Curie 1937). Lise Meitner and her colleagues discovered the exponential law governing the characteristic absorption of beta rays in matter (Taton 1964). Brooks and Rutherford conducted the earliest measurements of the penetrating power of alpha particles from various radioactive sources (Weeks 1956). Joliot-Curie and her husband, Frédéric Joliot, were the first to observe radioactive decay in which positrons were emitted, and the first to create a source of positrons in the laboratory.

The new measurements required new instruments and techniques, and women turned their talents in these directions as well. Marie Curie constructed early ionization chambers and invented the chemical process of fractionation—fractional crystallization—for extracting radioactive material from ore. She used this technique to discover polonium. Curie could have patented the technique, but she chose to regard it as a research tool and to forgo a personal profit (Taton 1964; Schwartz and Bishop 1958; Rhodes 1986). Edith Quimby, an American biophysicist and radiologist, improved methods of radiation dosimetry (Taton 1964).

The early research on radioactivity clearly opened a new and exciting field of science, but most crucial to the Manhattan Project was a sequence of specific discoveries that led directly to the work with nuclear fission that formed the basis for nuclear weapons. Five European women were closely connected with these discoveries. In outlining their roles, we are telling of the discovery of the physics underlying the atomic bomb.

## FIVE FEMALE SCIENTISTS AND THE DISCOVERY OF NUCLEAR ENERGY

The scientific tradition in which these women were educated was a formal one—not to say stodgy. Nineteenth-century physics, resting solidly on the mechanics of Galileo and Newton, seemed to be moving toward a complete understanding of the operation of natural forces. But chance discoveries of astounding new phenomena suddenly reopened the frontiers of physical science. Entirely fresh concepts were needed to understand these discoveries and their implications.

The revolution in physics broke out abruptly. Its start can be dated to a particular year, 1895, which saw the discovery of X-rays and radioactivity. The neutron was found in 1932, and nuclear fission was identified in 1938–39. In parallel with the experimental breakthroughs came the great theoretical achievements of relativity and quantum mechanics.

The German physicist Wilhelm Roentgen, who identified X-rays after noticing that something produced by his cathode-ray tube was darkening photographic plates, also found that a screen covered with a phosphorescent substance became luminous when placed near his experiments. The fact that X-rays produced marked effects on fluorescent and phosphorescent substances led Henri Becquerel to wonder whether phosphorescent or fluorescent natural materials themselves produced X-rays. He began to examine a collection of phosphorescent minerals he had inherited from his father.

It was purely luck that Becquerel began his search with uranium compounds. He might just as easily, or more easily, have selected zinc sulfide—an available, inexpensive material that exhibits phosphorescence—and the discovery of radioactivity, with all its implications, might have been delayed for fifty years (Bernal 1954). As it turned out, within four months of the discovery of X-rays Becquerel found that uranyl potassium sulfate emitted rays that, like the X-rays from cathode-ray tubes, affected a photographic plate through black paper and other opaque substances (Dampier 1949). The rays were produced spontaneously, however, without any apparatus at all. Energetic, penetrating radiation was being emitted from apparently inert, permanent chemicals, unstimulated by rays or light.

*Marie Curie*

Marie Sklowdowska Curie, a young Polish physicist studying at the Sorbonne in Paris, was searching for a topic for her doctoral research. She decided to follow up on Becquerel's discovery by trying to find materials that shared uranium's property of emitting radiation. In 1898—at the same time as a German scientist named Gerhard Schmidt—she succeeded in showing that thorium emitted similar radiation (Weeks 1956). She proposed calling all substances emitting Becquerel rays "radioactive."

Marie Curie persuaded her husband, Pierre Curie, to abandon his research on magnetism and crystal structure in order to join her in studying radioactivity. The Curies were the first to recognize that radioactivity was no mere laboratory curiosity but a fundamental property of certain atoms. Soon the Curies had found radioactive materials that exhibited much more intense radioactivity than either uranium or thorium. Working in an unheated shed, the Curies spent most of four years analyzing six tons of pitchblende, a uranium ore. By 1902 they had isolated a tenth of a gram of radium compounds, enough to determine the element's atomic weight and properties. The radium was so radioactive that it shone by itself in the dark and could inflict serious burns on individuals exposed to it (Parkinson 1985). In the course of their work, they discovered another new element, which they called polonium.

Becquerel had shown that "uranic rays," like X-rays, were able to cause the air around them to be more electrically conductive by increasing the number of ions. That effect intrigued Marie Curie, who decided in 1896 to look more closely into the matter. To measure the amount of ionization produced by the rays, she built an ionization chamber, in which the electrically charged ions produced by the passage of radiation were collected on metal electrodes. By 1897 she had built equipment that could measure the charge, using a quartz electrometer that had been developed in the course of her husband's recent discovery of piezoelectricity (the appearance of electrical charges on the surface of some nonconducting crystals when subjected to mechanical pressure). Marie Curie was able to measure the intensity of the radiation given off by a variety of uranium compounds, thus establishing a quantitative base for the study of radioactivity.

Marie Curie noted that the intensity of the radiation from a material was proportional to the quantity of uranium in the material, no matter what its chemical composition. This showed that the radiation came not from the compound as a whole but specifically from the uranium. Radioactivity, in other words, was a property of individual atoms of uranium and other elements (Taton 1964; Curie 1937), and was therefore an atomic phenomenon, not a molecular one.

In 1901, Pierre and Marie Curie, working with Becquerel and Jean Baptiste Perrin, a French physicist and chemist, put forward the view that radioactive decay transformed a radioactive atom into an atom of a different element and released the potential energy contained in the original atom (Taton 1964). This was a great shock to the nineteenth-century belief in the immutability of elements. Matter was apparently changing from one element to another of its own accord (Bernal 1954). As if that were not enough of a puzzle, even microscopic amounts of radioactive materials gave off appreciable energy. Radium glowed in the dark. This implied that atoms contained energy in quantities enormously greater than the amounts that had been extracted from the energy sources used by human beings up until that time.

Marie Curie shared the 1903 Nobel Prize in Physics with her husband and Henri Becquerel for their discovery and characterization of radioactive elements and their properties. She was the first woman to be awarded a Nobel Prize. Becquerel received the honor for his discovery of the phenomenon of natural radioactivity, and the Curies for their ensuing study of radioactivity.

Curie's original and influential work did not stop with the discovery and description of radioactive elements and the invention of related techniques and equipment. She showed that the electric charges transported by beta radiation are negative, and she helped establish that beta radiation consists of high-speed electrons (Hellmans and Bunch 1988; Taton 1964). After Pierre's death in 1906, Marie took over his chair at the Sorbonne and directed the laboratory that had belonged to both of them. She was awarded a Nobel Prize in Chemistry in 1911 for additional work on radioactive elements (L. M. Jones 1990).

Marie Curie won two Nobel Prizes, founded and directed a major research laboratory, and produced an enormous amount of important scientific work. She became the symbol of women's ability to succeed in the physical sciences. Even so, Harvard and Princeton refused to present her with an honorary degree during her visit to the United States in 1921, and neither the French nor the American National Academy of Science would elect her a member. Yale did award her an honorary degree, its sixth ever to a woman.

In addition to her scientific work, Curie raised two daughters, one of whom would win a Nobel Prize of her own and the other of whom would marry a Nobel laureate. Curie's younger daughter, Eve, published a biography of her mother in 1937 that, though it certainly documented the scientific ability of this extraordinary woman, depicted her life as rather grim. According to Eve Curie's account, her mother lived only for her work until her death from pernicious anemia in 1934. The biography became the basis for future accounts that depicted Marie Curie as a self-sacrificing scientific super-woman.

More recent scholarship, however, has included facts that were well known to Curie's contemporaries, such as the scandal surrounding her affair with a married physicist, Paul Langevin, which nearly resulted in Curie's dismissal from the Sorbonne. In 1921, perceptions of her character still suffered to the extent that some wished to bar her from the United States. Other information omitted from her daughter's account includes Curie's wartime activities: She spent considerable time at or near the front during World War I with X-ray machines mounted on the back of automobiles known as "Little Curies." She slept in field hospitals near the front lines and rehabilitated her reputation in France by her heroic care for the wounded (Giroud 1986; Quinn 1995). She thoroughly enjoyed being with her children, but was glad to accept the help of her father-in-law in raising them. The recent work reveals Curie to have been passionate about everything she did, including science. As one Frenchwoman who knew Curie remarked, "She became Parisienne!"

*Mileva Marić*

While the Curies pursued the experimental study of radioactivity, theorists had not been idle. A major step toward the discovery and understanding of the huge energy release in nuclear fission was the concept of mass-energy equivalence, a prediction of the special theory of relativity published by Albert Einstein in 1905. Until recently, it was thought that Einstein had developed this concept and other major theoretical advances essentially alone. He worked without any close collaborators and earned a living as a clerk in the Swiss patent office. Since the publication in 1987 of the first volume of *The Collected Papers of Albert Einstein*, however, there has been considerable discussion of the role that Einstein's first wife, Mileva Marić, had in the development of his most important ideas, all of which he published during or shortly after their marriage (Trbuhović-Gjurić 1988; Troemel-Ploetz 1990).

Marić, a physicist of Yugoslavian descent, was four years older than Einstein. They met as students at the Zurich Swiss Federal Polytechnic, where she was the only woman in a course to prepare secondary physics and mathematics teachers (Holton 1994). She evidently met Einstein in required classes, and a friendship grew into a romance. Recent research suggests that she may have collaborated with her husband in a number of areas of physics, including the development of the theory of relativity, although her contributions were never acknowledged publicly

(E. Walker 1991; Krstic 1991). As more of Einstein's papers are published, we may learn more of his relationship with his first wife and of her role in science. There have been several surprises already, such as the fact that Mileva and he had an illegitimate daughter and the revelation that Einstein had an extramarital affair with the woman who would become his second wife.

What has been published so far shows that at the very least Marić played the role of a supportive scientific colleague. The letters that Einstein wrote during their several enforced separations discuss the details of his work. (Her letters, of which fewer have survived, are briefer and more dedicated to their romance.) On April 4, 1901, he wrote to Mileva that he and his friend Michele Besso had "discussed the fundamental separation of luminiferous ether and matter, the definition of absolute rest, molecular forces, surface phenomena, dissociation. He's interested in our research, though he often misses the big picture by worrying about petty things. This pettiness torments him with all sort of nervous ideas. The day before yesterday he went on my behalf to see his uncle, Prof. Jung—one of the most influential professors in Italy—to give him our paper" (Einstein and Marić 1992, p. 41).

The following month, Einstein wrote that "The local Prof. Weber is very nice to me and is interested in my work. I gave him our paper. If only we were fortunate enough to pursue this beautiful path together soon but fate seems to have something against us" (Einstein and Marić 1992, p. 52). The letters are an amusing combination of such mundane details as missing umbrellas and new lodgings, really gooey romantic prose, and fairly lengthy discussions of ideas in physics.

The letters do not tell us whether Mileva actually helped in the research. One recent writer cites analysis by four established (male) historians of science as indicating that Einstein's use of the work "our" was intended to cheer up Mileva, who, after bearing an illegitimate daughter, giving the baby up for adoption or losing it to illness, and failing her exams at the Polytechnic (not for the first time), was suffering from depression (Holton 1994). Mileva and Albert married and had two legitimate sons, and she seems to have given up any hope of a career for herself. Einstein felt increasingly trapped by domesticity, and Mileva became more introverted and less interesting to him as he began to have extramarital affairs. The marriage ended in 1919. Mileva remained in Zurich, where she died in 1948.

It's abundantly clear that Albert Einstein's most productive years as a physicist came during his marriage to Marić and shortly afterward. When Einstein won the Nobel Prize in 1921 he gave his prize money to Marić, even though they were divorced by then (Trbuhović-Gjurić 1988; Troemel-Ploetz 1990; Krstic 1991). In Marić, Einstein had a supportive companion in science who may have helped him materially in the research for which he became famous.

*Irène Joliot-Curie*

For some thirty years after the discovery of mass-energy equivalence, the scientific community refused to accept the possibility of a significant release of nuclear energy. By the mid-1930s, the three most original and widely recognized living physicists had argued against it. Einstein had compared the possibility of a practical use for nuclear energy to shooting in the dark at scarce birds, Rutherford had dismissed it as moonshine, and Bohr felt that the possibility became more remote with each new discovery about nuclear reactions (Rhodes 1986). Each of these men had made radical changes in the conceptual foundation of physics, but none of them was prepared to accept the practical implications of Marie Curie's experimental findings of the enormous release of energy in radioactive decay.

Irène Joliot-Curie worked in her mother's institute and extended her work. In 1924, with Catharine Chamié, she made the first determination of the half-life of radon, using an ionization chamber (Parkinson 1985). Working with her husband, who had been Marie Curie's assistant, Joliot-Curie noticed a highly penetrating electrically neutral radiation from the bombardment of beryllium by alpha particles. (The Joliot-Curies made their discovery independently, but it had been preceded by that of Bothe and Becker, two German scientists, in 1931.) The neutral penetrating radiation was initially thought to consist of gamma rays, but in 1932 Sir James Chadwick, an English physicist, identified it as composed of neutrons (Hellmans and Bunch 1988). With the discovery of the neutron, scientists had found all the constituents of the atomic nucleus.

Irène and Frédéric Joliot-Curie were the first to observe radioactive decay in which positrons were emitted, and apparently also the first to create a source of positrons in the laboratory. In 1934, they discovered artificially induced radioactivity, producing an unstable isotope of phosphorus by bombarding aluminum with alpha radiation. Their experiments gave the first chemical proof of artificial transmutation of elements. They had demonstrated that it was possible not only to chip pieces off nuclei, but also to create new nuclei, which had to release some of their energy in radioactive decay. In 1935 they shared the Nobel Prize in Chemistry for their discovery that radioactivity could be produced artificially, and for their synthesis of new radioactive elements (Walton 1989; Clark 1961; L. M. Jones 1990).

The Joliot-Curies had produced, in the laboratory, a radioactive isotope of an ordinary stable element. Until that time, the only radioactive elements known were the naturally occurring ones. Their subsequent discovery that nearly all elements become radioactive when bombarded with neutrons led to experiments in which different elements were irradiated by neutrons in a search for new radioactive isotopes. The leaders in these experiments were Enrico Fermi's group in Rome and Lise Meitner's in Berlin. The Joliot-Curie discovery also meant that natural radioactivity represented only a residuum of extremely long-lived radioactive atoms that had not yet had time to achieve stable states.

Until 1937, all changes in atomic nuclei that had been identified were produced by adding relatively small particles to nuclei or by causing small particles to be ejected from nuclei. The largest fragment ejected had been identified as an alpha particle, which contains two protons and two neutrons. When nuclei were exposed to neutron bombardment, they changed because alpha particles, protons, or neutrons were knocked out of them or neutrons were added to them. The possibility of fission, in which a nucleus would be split into two roughly equivalent pieces, each with many protons and neutrons, had never occurred to most workers in the field.

There is some evidence that Irène Joliot-Curie and her team may actually have been the first to detect fission, although they did not recognize it. In 1934, when they bombarded uranium with neutrons, a nuclide was produced that may have been the fission product barium. At the time, it was interpreted as a transuranic element (Walton 1989).

*Ida Noddack*

Ida Noddack, a German chemist, first suggested the possibility of nuclear fission in the early 1930s (Starke 1979; Noddack 1934; Walton 1989). In 1934, Fermi bombarded uranium with neutrons in his laboratory in Rome and identified a new type of radioactivity—one whose atomic chemistry did not fit with elements on the periodic table close to uranium. He viewed this radioactivity as evidence of a new transuranic element and eagerly published his results. Ida Noddack, codiscoverer of the element rhenium, immediately published a paper questioning Fermi's chemical identification of the radionuclide as a transuranic element. She proposed that the radioactivity came from the splitting of uranium nuclei rather than from the buildup of transuranic nuclei (Jungk 1960; L. M. Jones 1990).

"It would be equally possible to assume that when a nucleus is demolished in this way by neutrons, nuclear reactions occur which may differ considerably from those hitherto observed, ... " Noddack wrote in her landmark paper. "It would be conceivable that when heavy nuclei are bombarded with neutrons, the nuclei in question might break into a number of larger pieces that would no doubt be isotopes of known elements but not neighbors of the elements subject to radiation" (Noddack 1934).

Fermi greeted this idea with derision, as did Otto Hahn, who was subsequently awarded the Nobel Prize in Chemistry for the discovery of fission. The nuclear binding energies that held protons and neutrons together, Fermi argued, were already known to be in the range of millions of electron volts, while his bombarding neutrons had had far less energy—not enough, he believed, to allow the process being suggested by Noddack to take place. Fermi's reaction may also have been colored by the fact that the creation of new transuranic atoms would have been a thrilling accomplishment, possibly of Nobel Prize quality; the formation of known elements by some previously unobserved process was a much less appealing interpretation.

Noddack reported that a year or so later, "my husband [Walter Noddack] suggested to Hahn, by word of mouth, that he should at least make some reference, in his lectures and publications, to my criticism of Fermi's experiments. Hahn answered that he did not wish to make me look ridiculous as my assumption of the bursting of the uranium nucleus into larger fragments was really absurd" (Jungk 1960, p. 62). Despite Noddack's distinguished research record, the scientific establishment ignored her hypothesis (L. M. Jones 1990). Four years later, the work of Meitner, Hahn, Strassman, and Frisch proved Noddack's interpretation correct (Walton 1989; Spradley 1989; Starke 1979).

*Lise Meitner*

Lise Meitner was very active in research on radioactivity, and she played the senior and leading role in most of her collaborations in this area. With Otto Hahn, she discovered the first example of nuclear isomerism in 1921 (Hampel 1968; Watkins 1984). It was Meitner, already recognized internationally for scientific studies, whose work established the existence of fission.

In the autumn of 1934, Meitner began experiments on the interaction of neutrons with uranium, studies similar to those of Enrico Fermi in Rome. While bombarding uranium with neutrons, Meitner and Hahn detected the unexpected presence of the element barium in the reaction products. In retrospect, it is clear that they, like Fermi and possibly Joliot-Curie, had observed nuclear fission, but this was not understood until 1939.

By 1938, Meitner, Hahn, and a young chemist named Fritz Strassman had found at least ten new radioactive substances in uranium bombarded by neutrons. In March of that year, however, Germany's incorporation of Austria subjected Meitner, an Austrian Jew, to German law. In July, friends persuaded her to leave Germany for Sweden. Hahn and Strassman continued their work on uranium.

In December 1938, Hahn and Strassman, using new chemical techniques, identified radioactive barium in the bombarded uranium. They immediately contacted Meitner, who had been the leader of their group, for her help in interpreting this puzzling result (Rhodes 1986; Sime 1996).

Like most scientists at the time, the Berlin group had expected the bombardment of uranium with slow neutrons to produce the nuclei of other heavy elements, possibly including radium. After bombarding the uranium, they dissolved it and separated new elements chemically. In the separation, they added a barium compound as a carrier, as barium is chemically similar to radium and precipitation of the stable barium was expected to precipitate any radium atoms in the sample as well. When they precipitated the barium, some of the radioactivity came with it, as they had hoped. When they tried to separate the radioactivity from the barium, however—which would have established the existence of an important process called double-alpha-particle emission—they failed. The radioactivity was not coming from radium. All the chemical tests indicated that it was, in fact, coming from the barium (Vare and Ptacek 1988).

Lise Meitner spent Christmas 1938 with her nephew, Otto Frisch. Excited by the news of Hahn and Strassman's results, they discussed the problem of barium in the bombarded uranium. Meitner proposed that the radioactive barium came from a process in which a heavy atomic nucleus broke into two parts of roughly equal size. The nuclear fragments, being unstable, were themselves radioactive. Meitner and Frisch calculated that this newly discovered process would release enormous amounts of energy. They named the process "fission," after the term for the splitting of a biological cell, and prepared their results for publication as a letter to the British journal *Nature* (Sime 1996).

The letter to *Nature* contains the following passage:

> It seems therefore possible that the uranium nucleus has only small stability of form, and may, after neutron capture, divide itself into two nuclei of roughly equal size (the precise ratio of sizes depending on finer structural features and perhaps partly on chance). These two nuclei will repel eachother and should gain a total kinetic energy of c. 200 MeV, as calculated from nuclear radius and charge. This amount of energy may actually be expected to be available from the difference in packing fraction between uranium and the elements in the middle of the periodic system. The whole "fission" process can thus be described in an essentially classical way. (Meitner and Frisch 1939)

Meitner was the first to grasp the significance of the new experimental results by Hahn and Strassmann. Although she had both initiated the series of experiments that produced the findings and provided the theoretical interpretation that made their importance clear, Hahn refused to share credit for the discovery and broke off their close collegial relationship. Fear of the Nazis may have been a factor at the time, but even after the war he refused to acknowledge Meitner's role in their joint work when he and Strassman shared a Nobel Prize for the discovery of nuclear fission (Sime 1996).

Although she was not happy working in Stockholm, Meitner refused to emigrate to the United States during the war because she did not wish to work on weapons and wanted to be near her friends and family in Europe. In 1943 she was invited to Los Alamos as part of the British delegation, but refused. "I will have nothing to do with a bomb," she wrote (Sime 1996, p. 304–307). Meitner's expertise and the respect in which she was held by the physics community make it likely that had she come to the United States she would have played a prominent role in the Manhattan Project. Even in Stockholm, she remained in contact with both the German and the Allied scientific communities, and she must have been aware of the uranium projects.

Nor were the other founding mothers of nuclear physics available to the U.S. bomb effort. Marie Curie had died in 1934, and Irène Joliot-Curie elected to remain in Paris with her husband during the war. As Frédéric became increasingly involved with the Resistance, Irène maintained the operation of the research laboratory. Reluctant to believe the extent of Nazi ambitions, they continued to publish results on nuclear fission long after Allied scientists had adopted voluntary censorship to keep nuclear information out of Nazi hands.

In 1939, the Paris group demonstrated that the fission of uranium nuclei could lead to a chain reaction (Hellmans and Bunch 1988). Physicists had already recognized that when uranium nuclei split, they necessarily liberated several neutrons, because heavy nuclei carry a much larger proportion of neutrons than light nuclei do. If each fission of a nucleus not only released energy but also caused other nuclei to fission, and if these nuclei released further energy and caused additional fissions, and so on, the event would constitute a sudden, huge release of energy—in other words, an explosion. (Controlling the flux of neutrons in the material so that each fission in effect produced just one new fission would mean that a nuclear reactor could be developed.) When the Paris group published its work, scientific leaders in the United States and Britain feared that German physicists might recognize its military potential, but there wasn't much to be done about it.

Security in the world of open research continued to depend on voluntary restraint by researchers. In 1942, Gertrude Scharff Goldhaber, working at the University of Illinois because her German birth barred her from the Manhattan Project, discovered that spontaneous nuclear fission produced neutrons—a discovery with huge implications for the possibility of predetonation of a weapon. Her discovery remained secret until the end of the war, despite the fact that neither Scharff Goldhaber nor her husband, an Austrian nuclear physicist, had military clearance. Scharff Goldhaber would later be the first female Ph.D. hired by Brookhaven National Laboratory and the third female physicist elected a member of the National Academy of Sciences (Bond and Chasman 1998).

Meitner and Joliot-Curie remained in Europe. Noddack was German, and Marić, who had been born in Yugoslavia and held Swiss citizenship, had dropped out of science. Thus the women who had pioneered nuclear research were not available to the Manhattan Project. In the United States, the new field of nuclear physics had relatively few practitioners. Many scientists were already engaged in war-related research, particularly on the radar project at MIT. European emigrants therefore played leadership roles in the race to develop the bomb. Had Joliot-Curie and Meitner joined them, these women would surely have been among the project leaders. Though this didn't happen, it was they who had laid the groundwork for the research and development efforts of the Manhattan Project.

# 3  The Physicists

WORLD WAR II was a physicists' war. Its dominant new military technologies, radar and the atomic bomb, depended on research in physics, and war-related research demanded ever-increasing numbers of physicists, a small community at best. In 1939, the American Physical Society had 3,600 members. By the end of 1941, some 1,700 physicists, about a quarter of the profession, were employed in defense-related research. This group included an even higher percentage of professional leaders and those working in such emerging fields as nuclear physics. In the summer of 1943, the defense projects still urgently needed 315 physicists for high-priority research (Kevles 1978).

Female physicists were then, as they are now, a small minority of the profession. In the 1930s, women earned about 3 percent of the doctorates awarded in physics in the United States. They worked primarily in women's colleges and other academic institutions that provided limited research opportunities. A very few joined industry (Kevles 1978). Probably 100 women, at most, were active in physics research at the outbreak of World War II.

One of these women, Elda "Andy" Anderson, had received a Ph.D. from the University of Wisconsin in atomic spectroscopy. She held a teaching position at Downer's College in Milwaukee. The Cyclotron Group at Los Alamos persuaded her to take a leave of absence from her teaching duties and move west. She worked on studies of the fission process, including measurements of such basic parameters as the number of neutrons produced per fission and the time delay, if any, before the emission of neutrons.

Anderson, fifty years old with white hair, seemed a senior citizen compared with most of the workers at Los Alamos. She had a reputation as an eager and cooperative worker. Other women remember that she lived in the dormitory and worked mostly at night, clad in jeans and a plaid shirt—a daring fashion statement for the times. Among other work, Anderson prepared the first sample of nearly pure uranium-235 received by the Los Alamos group for use in the experiments (Barschall 1991).

Rose Camille LeDieu Mooney-Slater, a physicist and crystallographer, studied the structures of crystals and crystalline materials with X-ray diffraction at the University of Chicago's Metallurgical Laboratory—the Met Lab. She was born in New Orleans in 1902 and studied physics at Newcomb College of Tulane University, where she received her B.A. and master's degrees. She earned her Ph.D. in physics from the University of Chicago in 1932, then returned to Newcomb as a professor of physics; students describe her as an inspiring teacher. She remained affiliated with Newcomb through the war, becoming full professor and head of the Department of Physics in 1941.

In 1943–44, Mooney-Slater took a leave of absence from Newcomb to join the Manhattan Project at the University of Chicago, specifically to establish an X-ray crystallographic laboratory with W. H. Zachariasen, a noted crystallographer. She became associate chief of the X-ray Structure Section of the Met Lab, and thus one of the highest-ranking women on the Manhattan Project. After the war, she became a research physicist at the National Bureau of Standards (1952–56), the Massachusetts Institute of Technology (1956–81), and the University of Florida (1966–74). She married John Clarke Slater, a theoretical physicist and MIT professor, in 1954 (her second marriage). She received numerous awards for her achievements as a research crystallographer. She died in 1981 (*American Men of Science*; Moulton 1982; Catlett 1987).

Despite such women as Anderson and Mooney-Slater, the public and many male physicists refused to believe that women could handle physics. This view was surely a big factor behind the feeling of many women active in research that they had to remain single in order to dedicate themselves to science. Anderson never married, and Mooney-Slater married only after establishing her reputation as a crystallographer. All twenty-three female scientists listed in the 1923 edition of *American Men of Science* were unmarried. (The word "Women" wasn't added to the title until 1971.) So were 615 of the 687 women in the 1927 edition (Pressey, Rossiter and Bromley, quoted in Kevles 1978, footnote 4, p. 205). Married female physicists were generally treated as talented amateurs, and were given lab space only if their husbands demanded it. The plight of female scientists during these years—and the strides that the war helped them make—is well illustrated by the case of Maria Goeppert Mayer.

Goeppert studied quantum mechanics with Max Born at Göttingen, the hotbed of German theoretical physics. In 1930, while completing her Ph.D., Goeppert married Joseph Mayer, an American postdoctoral fellow working in physical chemistry. They moved to Baltimore, where Joseph Mayer had received an appointment as associate professor of chemistry at Johns Hopkins.

Although Maria Mayer's training in the new field of quantum mechanics surpassed that of anyone else on the university faculty, she became a volunteer associate of the university. Her tiny salary came from helping a professor with his German. She remained active in research, however—even continuing to work with Born during summer visits to Göttingen until 1937—and she collaborated on a text on statistical mechanics with her husband.

In 1939, Joseph Mayer moved to Columbia University, where his wife was offered an unpaid lectureship without a laboratory. In late 1941, Sarah Lawrence College in Bronxville, New York, hired her to fill a half-time position teaching physics and physical chemistry. Finally, in 1942, Harold Urey, head of the Manhattan Project's gaseous-diffusion project, recruited her to work half-time at Columbia, giving Mayer her first real salary for physics research. Although she was nominally a part-time worker, she quickly became the leader of a group of twenty scientists and technicians. She later visited Los Alamos several times to work with Edward Teller, and she continued to teach part-time at Sarah Lawrence.

## The Met Lab's Atomic Piles

In nuclear fission, a uranium nucleus absorbs a neutron and then splits into two lighter nuclei. Each fission typically releases neutrons and a substantial amount of energy in the form of heat, electromagnetic radiation, and radioactive fragments. A nuclear chain reaction occurs when a neutron produced in one fission reaction triggers another reaction, which releases further neutrons that intercept other nuclei and continue the process. To produce a nuclear explosion, the fissioning material must be assembled so that each fission leads to more than one new fission. As the number of fission reactions increases, energy release grows explosively until the fissioning mass blows itself apart, stopping the chain reaction.

In 1939, physicists did not know whether a nuclear chain reaction was possible. Early experiments had shown that each fission released more than one neutron, that slow neutrons were more effective in causing fission than fast ones, and that only the rare isotope uranium-235 fissioned with slow neutrons. A group based at Columbia University, led by the Italian Nobel laureate Enrico Fermi and the Hungarian scientist Leo Szilard, began experiments to determine whether a nuclear chain reaction could sustain itself.

Since it was known that slow neutrons are much more effective in causing fission, the scientists needed to slow the neutrons down. Fermi and Szilard first used water around the uranium to slow the neutrons, but the water absorbed too many of them to allow a sustained chain reaction. Fermi lost interest, but the tireless Szilard found a source of very pure graphite to use in place of the water, and results became more promising.

Initially the physicists built atomic piles: Uranium was packed into holes in graphite bricks, and these bricks were enclosed in structures made of more graphite bricks. Instruments in the piles monitored the flow of neutrons. The sign of a chain reaction would be a growing, sustained neutron flux. In the spring of 1942, the experiments moved to the Metallurgical Laboratory at the University of Chicago. The first self-sustaining nuclear chain reaction was produced on December 2, 1942, in an old squash court under the football stands at Stagg Field.

Leona Woods, a graduate student at the University of Chicago, was one of the youngest members of Enrico Fermi's staff. Woods, who was completing her Ph.D. in molecular spectroscopy while the Manhattan Project experiments were taking place, stopped one day to talk to a physicist smelting metal bricks in a hallway. (He later became her first husband.) Woods's son, John Marshall III, describes the encounter: "She said to him, 'You guys have discovered how the chain reaction works!' The next day, Fermi said that she should be working with them. The [Manhattan] project was so secret they had to figure out something to do with a clever woman" (Folkart 1986).

Woods constructed the boron trifluoride neutron detectors used to gauge the flow of neutrons in the experimental atomic piles. She operated the electronic equipment and monitored the counters, and she was calling out the counts on the memorable day when Chicago Pile Number One first sustained a chain reaction. After operating for about half an hour at about 200 watts of power, the pile was shut down.

To celebrate the success of the experiment, Fermi was handed a bottle of Chianti, and paper cups were handed round to the exhausted crew. Woods broke the silence with the comment: "Let's hope that we are the first to succeed" (Seidel 1992). Woods and the other scientists present signed the label of the Chianti bottle, which was kept as a memento of the successful experiment.

Other women who worked on the Met Lab piles included Nancy Wood, a physicist who built excellent particle detectors. Wood's counters were in such demand by other physicists that she continued to construct them in Chicago after the war. Eleanor Gish and her colleagues at the Met Lab studied the important question of how many neutrons emerge during an average fission reaction, and how many of these can be expected to trigger the fission of a second nucleus (Kathren et al. 1994).

Patricia D. Walsh worked on instruments that could measure the radioactivity, weight, and conductivity of extremely small samples of transuranic elements. Born in Washington, D.C., in 1920, Walsh earned an A.B. in physics from Smith College in 1942. After working as an electronics instructor at the University of Chicago, she joined the Met Lab in October 1943 as a research assistant (Kathren et al. 1994; Seaborg 1977).

In January 1945, reactor construction and development was moved to Oak Ridge, Tennessee, and Hanford, Washington, and the Met Lab staff spent more time on less urgent theoretical work. Though the research environment altered in some minor ways—management cut the number of telephones, for instance—Walsh continued to work at the lab. She was involved in a number of memorable events that month—none of them, unfortunately, worthy of a Chianti celebration. On January 18, security officers chastised a number of staff members, including Walsh, for such violations as leaving reports out and not locking their files. On January 25, 1945, a radioactive spill involving beta and gamma emitters occurred in a room occupied by Walsh and three other staff members. Though the spill affected a wide area, reported radiation levels were not dangerously high. Walsh continued to work at the lab until the end of the war, when she joined the newly established Argonne National Laboratory. She retired in 1989 (Kathren et al. 1994).

Lorraine Golden studied biochemistry at the University of Chicago, receiving her B.S. in 1942. In 1943, she quit her job as an electronics instructor at the university and began working as a research assistant at the Met Lab. In October 1944, she married John A. Crawford (her second marriage), who also worked in the Instruments and Physical Measurements Group as a junior physicist and research assistant. Their City Hall wedding, typical of the romances of the Manhattan Project, was celebrated by the lab staff. Golden was involved in the same minor security violations and radiation spill as Walsh. She left the lab shortly thereafter (Kathren et al. 1994; Seaborg 1977).

## Producing Fissionable Materials: The Piles at Clinton and Hanford

More than 99 percent of natural uranium consists of the isotope U-238, which cannot sustain a chain reaction; the only isotope that can sustain one is U-235. Isotopes of a single element are chemically identical, differing only in the structure of their nuclei, so separating isotopes must be done by physical, not chemical, techniques. Doing this on a large scale promised to be extremely difficult. Manhattan Project scientists began working on industrial-scale processes for separating the isotopes even while they continued to study the basic physics of nuclear fission—sort of like constructing an automobile assembly line while the gasoline engine was still in development. Four methods of separating the isotopes were studied: gaseous diffusion, electromagnetic separation, thermal diffusion, and centrifugal separation.

Gaseous diffusion separates isotopes by allowing gas containing uranium to flow through many porous membranes. The gas molecules containing uranium-235 are a bit lighter, and therefore move a bit faster, than those containing U-238. After passing through thousands of membranes, the resulting gas contains more uranium-235 than natural uranium.

The only stable gas containing uranium proved to be uranium hexafluoride, which was not only unstudied but also toxic and highly corrosive. At Columbia, Maria Mayer conducted theoretical studies of the thermodynamic properties of uranium hexafluoride gas. She also investigated the possibility of using photochemical reactions for isotope separation, but her results didn't justify starting an experimental program on this separation method (Dash 1973; Rempel 1993; Sachs 1979).

At the Naval Research Laboratory in Washington, D.C., and at the Philadelphia Naval Yard, Philip Abelson worked on developing thermal-diffusion columns with a hot inner core and a cool outside. Liquid uranium hexafluoride in the columns flowed up the inner side and down the outer. Liquid enriched with lighter uranium-235 collected near the top of the column.

A team based at the University of Virginia attempted to use centrifuges to separate isotopes, but it did not show the promise of the other three techniques.

At Berkeley, the Nobel laureate Ernest O. Lawrence used his invention, the cyclotron—developed to speed up protons, crash them into other particles, and study the structure of the nucleus—to accelerate uranium ions through a magnetic field in arcs. The lighter uranium-235 ions curved just a little more than the uranium-238 ions. The Berkeley team also used the accelerators to produce neptunium and plutonium, the first elements ever discovered that were heavier than uranium. The discovery of plutonium in January 1941 had immense implications for the Manhattan Project because, like uranium-235, it was capable of fission. Plutonium could be produced in atomic piles, meaning that there there were now two paths to an atomic bomb: (1) separate the chemically identical isotopes of uranium or (2) supply plutonium from the fuel used in atomic piles. General Leslie Groves, who was in charge of the entire Manhattan Project, decided to pursue all possible paths. He simultaneously stepped up research on isotope-separation processes and began the design of huge atomic piles to produce plutonium.

To control the chain reaction in the first atomic pile, Enrico Fermi used rods of a neutron-absorbing material such as cadmium. When the rods were inserted into the pile, they stopped the fission chain reaction. To start the reaction, the rods were slowly pulled out of the pile while the neutron counts were carefully watched. To prevent a runaway reaction, which probably would have given all those present a lethal dose of radiation, a team of graduate students, known as the suicide squad, stood just above the pile with hammers and glass bottles filled with a cadmium solution. In the event of a disaster, they were to break the bottles and flood the pile. Clearly, the pile belonged somewhere other than the middle of a major city.

Fermi's team rebuilt the pile in the Argonne Woods outside Chicago. Leona Woods, who in July 1943 had married the physicist John Marshall, followed the project to Argonne. She continued to conduct experiments on its operation and the neutrons it produced until the birth of her first son in 1944. She hid her pregnancy under overalls and a denim jacket and worked until two days before the baby's birth.

Major construction of new atomic piles moved to Oak Ridge, there were fewer people and more room. But even this site proved too small to safely contain the plutonium production piles—which would today be called nuclear reactors. Groves ordered construction begun at Hanford, Washington—razing the town and displacing the few inhabitants—and hired E. I. Du Pont de Nemours to build the facility. Du Pont began designing the production reactors even before Fermi's group demonstrated that a chain reaction was possible.

After her baby had been born, Leona Marshall moved to Hanford to join her husband, a senior member of Fermi's team, in overseeing the operation and construction of the plutonium production reactors there. Her expertise in pile design and operation made her an ideal candidate for the position. The Marshalls were involved in the design of every nuclear reactor built during World War II. Fortunately, Leona's mother was willing to accompany her to Washington to help care for the baby. For the remainder of the war, Marshall devoted herself to the production of plutonium for the new weapon (Libby 1979). Colleagues remember her actively at work and not turning a hair at the profanity of the construction crews.

Katherine "Kay" Way attended Vassar College and Columbia University, where she studied mathematics with physics on the side. She read *The ABC of Atoms* by Bertrand Russell and decided that her primary interest lay in physics. She attended the University of North Carolina, where she became the first graduate student of John Wheeler, a brilliant young theoretical physicist who had worked with Niels Bohr. Even after completing the requirements for her doctorate, Way kept her position as a graduate student because jobs were scarce. In 1938, she received a Huff Research Fellowship at Bryn Mawr, which allowed her to graduate and continue her research.

In 1939, Way was appointed an assistant professor at the University of Tennessee. In 1942, while working on construction of a neutron source to produce neptunium-239, she moved to Washington to work on underwater warfare in a group led by John Bardeen, who later was one of the inventors of the transistor and won two Nobel Prizes in Physics.

Hearing rumors of a nuclear project at Chicago, Way called Wheeler and asked whether there was work for her. Because the Manhattan Project urgently needed nuclear physicists, she managed to arrange a transfer. In Chicago, she joined Alvin Weinberg in analyzing Fermi's data from the early atomic piles to see whether the flux of neutrons in the pile could be made large enough to permit a self-sustaining chain reaction. Their calculations were used in the construction of CP-1—the pile in which the first such reaction took place. She also worked on projects dealing with the "poisoning" of the atomic piles by fission products, which absorbed neutrons and shut down the reaction; the determination of reactor constants; and the organization of radioactivity data, which led to the Way-Wigner formula for fission-product decay.

Way commuted between Chicago and Tennessee in an ancient car that she had bought from a friend for $150. The production reactors at Hanford used her designs, and she visited the site. She knew of the atomic-bomb project and of the work at Los Alamos. She visited Los Alamos to "get acquainted," but did not remain there for an extended period (Way 1991). In the early summer of 1945, Way moved to Oak Ridge to continue work on nuclear decay and neutron cross-sections. She was deeply shocked by the destructive power of the atomic bomb at Hiroshima, and she began to question the use of nuclear weapons (Artna-Cohen et al. 1993).

Jane Hamilton was born in Denver in 1923. She attended the University of Chicago, where she received a B.S. in 1937, an M.S. in 1938, and a Ph.D. in physics in 1942 (Sylves 1987). While there, she married a fellow student, David Hall. The couple worked briefly as instructors of physics at the University of Denver, and then, in 1943, joined the Physics Division of the Met Lab. In 1944, Du Pont hired her as a senior supervisor at the Hanford Engineer Works (Kathren et al. 1994; Sylves 1987).

"We stayed at Chicago until early in 1944," David Hall later said, "and then we went out to Hanford to babysit the construction of three reactors. We reached Hanford in June or July. I remember when we first got there the houses weren't ready and we stayed in dormitories. Blast! Those bedsheets were hot. You touched them and they were hot. I had never experienced that dry heat before. Our front lawn, after we got a house, had wild asparagus coming up. The only remarkable thing about our house was that the contractor apparently had not been able to get regular bathtubs and so the bathtubs were poured concrete. Kind of gritty on your bottom" (Sanger 1989, p. 131).

The Manhattan Project's contract with Du Pont for operating Hanford specifically excluded the conduct of research. There were many scientists stationed at Hanford to work on the technology and improve the operational program, but this counted as development rather than research. According to Herbert Parker, section chief of Health Physics: "In our particular area in radiation protection, although we were excluded from doing research, there were a few things that we so compellingly felt we wanted to get an improved answer on that we had a category described as 'special studies.' If you had access to all the literature of the first two years here, you'd find that there was a Special Studies section that was headed by Jane Hall a good share of the time. I was lucky enough to get Jane into that group."

According to Parker, his luck in acquiring Jane Hall arose from nepotism: "Jane Hall was the wife of Dave Hall. They were both physicists, and by the rules of the game, we could not allow them to work in the same division which is how I came to get Jane, fortunately."

Parker first assigned Jane Hall the task of assessing the safety of the production reactors. In the course of her work, she investigated and reported on the hazards of plutonium inhalation (Hanford Works-HW-14058 1945; Hall 1944). One piece of equipment she worked with consisted of two copper spheres with a Geiger counter inside each. One sphere was twice the diameter of the other, and therefore would contain eight times as much radioactive liquid. This apparatus, like much of the equipment at the lab, was referred to by a proper name. According to Parker:

> It was called Devizes after a well-known English limerick.... Devizes is an English resort, a seaport. The limerick is:
>
>> There once was a man of Devizes
>> Whose balls were different sizes
>> The one was so small
>> It was no use at all
>> The other had won several prizes.
>
> We enjoyed this name tremendously for this apparatus.... And then we got into this business where Jane Hall became head of our special studies thing and soon discovered that all our instrument names had some meaning. And she insisted on knowing the meaning of this thing.
>
> It's surprising how quickly you'll learn to modify the old English limericks, to say:
>
>> There once was a man of Devizes
>> Whose eyeballs were different sizes....

Of course, nowadays, you wouldn't bother to change it. The girls would not be shocked by the original.... I don't think Jane was significantly fooled for very long.

In 1945 Jane Hall joined the staff at Los Alamos, and she worked there with the Manhattan Engineer District in 1945 and 1946.

Perhaps the most important insight into the operation of the Hanford reactors came from a female physicist who never visited Hanford. After construction was completed on the first large-scale plutonium production reactor in 1942, the reactor operated for only a few hours before shutting itself down. The chain reaction had stopped. Fermi suspected that some fission product being produced in the reactor was bringing the chain reaction to a halt by capturing large numbers of neutrons. "Ask Miss Wu," Fermi was reportedly told.

Chien-Shiung Wu had traveled to Berkeley from her home near Shanghai, China, in 1936 to pursue graduate studies in physics at the University of Michigan. Meeting her future husband and seeing the Berkeley labs changed her plans, and she received her Ph.D. in 1940 for a dissertation completed under the direction of Ernest Lawrence. Her research topic was a study of nuclear interactions of noble gases. Fellow graduate students describe her as extremely competent and very independent. If offered help with an experiment, she would announce that she preferred to do it herself (Freedman 1991).

When Fermi called Wu, she was able to diagnose the problem as the presence of Xe-137, a xenon isotope that was extremely efficient in capturing neutrons. Once Wu had made the diagnosis, the problem was resolved—thanks to the foresight of an engineer known locally as "Uncle George," who had insisted on providing excess core capacity within the reactor when it was built. "I don't care what those long-hairs say, add 10 percent more core to the pile," the legend has this practical engineer remarking. Fermi now used this extra space for more uranium fuel rods, providing enough additional neutrons to compensate for those being absorbed by the xenon.

Chien-Shiung Wu married Luke Chia Liu Yuan, a fellow Chinese physicist at Berkeley. She took teaching jobs at Smith College and Princeton University. Finally she joined Harold Urey's group, which was trying to separate U-235 from U-238 by gaseous diffusion and worked in a laboratory set up in a converted automobile showroom near Columbia University. She helped to develop the uranium-enrichment process and build an improved gamma detector (Rossiter 1995; *Industrial Research* 1974; Lubkin 1971; Yost 1960).

## Designing an Atomic Bomb

In addition to obtaining several kilograms of either uranium-235 or plutonium, the amount needed for one bomb, the Manhattan Project had to find a way of putting a "supercritical" mass together fast enough so that the bomb would release its energy before destroying itself. For an explosion of the necessary magnitude to take place, the chain reaction had to last for about fifty-seven generations (each of which takes about 0.00000001 seconds), with the number of fissioning nuclei growing exponentially at each generation. Energy is released with each fission in a nuclear explosion, but the bulk of it is not released until the last few generations, when by far the largest number of fissions occur. The problem for the Manhattan Project scientists was that the chain reaction began as soon as the two parts of the mass of uranium or plutonium approached each other. Unless the entire supercritical mass could be assembled extremely quickly, there wouldn't be enough neutrons available to keep the reaction going for the necessary number of generations, and the result would be too little, too soon: The bomb might melt or splatter, or there might be a small explosion, but nothing like the huge energy release characteristic of a nuclear weapon. An explosion equivalent to "only" ten tons of TNT, say, instead of 20,000 tons, was called a "fizzle."

A supercritical mass of uranium-235, the scientists found, could be assembled quickly enough by shooting a plug of uranium into a uranium sphere with a hole in it. The fissioning material had to be surrounded by a material that reflected neutrons, and the bomb required an initiator—a device to provide neutrons to get the fission reaction going. The first uranium bomb was detonated over Hiroshima.

A supercritical mass of plutonium could not be assembled in the same manner because its isotope—plutonium-240—spontaneously fissioned at too high a rate, meaning that the fission reaction would begin, and end, too soon. Los Alamos scientists eventually learned to make shaped explosive charges that compressed a sphere of plutonium metal extremely quickly. Plutonium bombs were tested at Trinity and used over Nagasaki.

The physicists involved with the Manhattan Project had a general idea, of course, of the immense power of the weapon they were creating, but they needed to find some way of predicting exactly how it would release its energy. While working on isotope separation at Columbia, Maria Mayer, with Edward Teller's encouragement, attempted to calculate some of the effects of a nuclear-fission explosion. Her calculations showed that electromagnetic radiation made up a surprisingly large fraction of the total energy released. She calculated the opacity of uranium at very high temperatures and how the fissioning mass absorbed the radiation. Her results were considered interesting but unimportant at the time, but they eventually provided the basis for the design of the hydrogen bomb. Her association with Teller continued later at Chicago, where she began to study the origin of the elements. This led to her recognition of nuclear magic numbers—configurations of neutrons and protons that are unusually stable—and the development of the shell model of the nucleus, for which she would share the Nobel Prize in 1963.

Because she was of German extraction, Mayer did not receive full security clearance even though she had become a U.S. citizen. Teller recounts a trip with Mayer to Washington to consult with a bureaucrat about her opacity calculations. Mayer pointed out that in order to do the calculations she would need to know the temperature range on which the military needed data. The bureaucrat provided it, and she gasped at the extremely high temperatures involved, since she had not been cleared to know about the bomb. In the same discussion, Mayer asked what she should tell the students working with her on the project, since even the word "uranium" was classified at the time. The bureaucrat suggested that she simply tell them that they were working with element number 92 (Teller 1991).

## Female Physicists at Los Alamos

General Groves consolidated bomb-design efforts at Los Alamos, New Mexico, in the spring of 1943. This had the joint advantages of bringing all the experts to one site and of making it easier to maintain security on critical details of weapon design. In an inspired personnel decision, Groves appointed the theoretical physicist J. Robert Oppenheimer director of the laboratory, which would actually be a military base. Because physics research was concentrated at Los Alamos, female physicists also concentrated there. Like their male colleagues, women were involved with both theory and experiment, but the majority were in experimental work.

At the time of the Manhattan Project, nuclear physics had just begun to develop as a science. Most of its practitioners in the United States, male and female, were relatively young—the majority of them in graduate school or just out of it. They arrived at the secret laboratory with a variety of expectations.

Several had followed their husbands west. Elaine Sammel went to college, but her first love was dancing; her résumé included a stint with the Chicago Opera Ballet. Her father allowed her to join a USO tour on the condition that she complete her bachelor's degree when she got home. She got her B.S. in physics from the University of Chicago and took a job at the Met Lab in optics research. There she met Harry Palevsky, an electrical engineer working on instrumentation. When Palevsky was assigned to Los Alamos to work in the Electronics Group, he learned that the lab also needed good people in optics. He suggested that Los Alamos hire Sammel, with whom he was already in love.

At Los Alamos, Sammel worked on the development of optical instrumentation while Palevsky worked on the trigger mechanism for the bomb. On a hot day in July 1945, they drove down the mountain to Santa Fe and were married. Elaine Palevsky did not pursue science in peacetime, concentrating on raising their five children. She also organized a ballet school near Brookhaven National Laboratory, where Palevsky worked after the war (Granados 1997).

Another young Chicago Ph.D., Elizabeth "Diz" Riddle Graves, moved to Los Alamos with her husband, Al, in 1943. They had met and married while graduate students in physics. Al Graves completed his Ph.D. and accepted a position on the faculty of the University of Texas at Austin. Elizabeth Graves received her doctorate in nuclear physics at the University of Chicago, but couldn't get a job at the University of Texas because of nepotism rules. Al Graves insisted that she be allowed to work at Los Alamos as a condition of his going there. If he hadn't insisted, she probably would have been recruited anyway.

Elizabeth Graves had worked on neutron-scattering experiments during her dissertation research, so at Los Alamos she began work on the selection of a neutron reflector to surround the core of the atomic bomb. The reflector kept neutrons inside the fissioning mass and speeded up the explosive growth of the chain reaction. No one had yet measured the effectiveness of various materials in scattering high-energy neutrons, but Elizabeth Graves was one of the few physicists in the country with experience in fast-neutron scattering and using a Cockroft-Walton accelerator, one of which had been brought to Los Alamos from the University of Illinois.

Colleagues report that Graves worked very hard and was very good at her job. Although her outlook was basically conventional, she could think independently and assert her point of view when it was necessary. She also had a sense of humor. She once bet colleagues that she could persuade a very proper European physicist to precede her through a door. She collected by telling him that she had ripped her dress and that modesty dictated that he go first.

Graves, like Leona Marshall, had a baby while working at Los Alamos. In fact, she finished a series of experiments while in labor, standing on the experimental floor, timing her contractions with a stopwatch (Barschall 1991).

Graves was pregnant when the first atomic bomb was tested at Trinity, and, concerned about the safety of the unborn child, she and her husband requested assignments well away from the test. They were assigned to monitor radiation that might drift from the test area, and they checked into a cabin in Harry Miller's Tourist Court in the town of Carrizozo, New Mexico, east of the test site. Their baggage included a seismograph, a Geiger counter, a shortwave radio, and a portable electric generator. A historian describes the scene:

> In Cabin Number 4 the Graveses spread out their equipment on the creaky double bed and told the inquiring owner that they were on a cross-country trip and would stay only two nights. The owner seemed more interested in where the couple had gotten gasoline to make such a trip than their weird instruments. In fact, Al Graves had frugally hoarded his ration stamps so he could drive from Los Alamos to Santa Fe every month for a cello lesson.
> ... In Carrizozo, Al Graves anxiously watched his wife puttering with the Geiger counter that rested on the window sill facing Trinity forty miles westward. Diz Graves was seven months pregnant, and Al worried that the strain of the last hours might injure her health. He put a steadying arm around her and drew her to him. Over the short-wave set they could hear Sam Allison conversing excitedly with the pilot of the B-29. The Graveses made a last check of their instruments. Then together they waited for the

jagged crests of the Oscuras to emerge from the night.... (Lamont 1965, pp. 193, 227)

During the countdown, the shortwave radio failed momentarily, and Diz stammered out the last seconds on her own. Because of their distance from the site, the windblown fallout from the test did not arrive until afternoon. At about 3 P.M. the couple began getting disturbing readings on their Geiger counter. At 4:20, the needle of the counter shot off the scale, and Al Graves phoned the base camp. The scientists and military discussed whether to evacuate the residents of Carrizozo, but decided against it. Within an hour the radioactive cloud had passed over the town and the counter readings had dropped (Lamont 1965).

Mary Langs traveled from her home in Detroit to the University of Washington intending to study medicine. She became interested in physics, however, and received her bachelor's degree in the field in 1939. She received a fellowship at Mills College in Oakland, California, and obtained a master's in physics in 1941. In 1942, while continuing her studies at George Washington University in Washington, D.C., she met and married Harold Argo, another graduate student in physics.

According to her husband, Mary found the atmosphere at George Washington very encouraging to female students. As staff members were drawn into war-related research projects, she became an instructor and taught freshman physics.

Both Mary and Harold were students of Edward Teller, whom they described as a considerate and supportive mentor, and they followed him to the Manhattan Project in 1944. They were sent to Los Alamos, where they worked under Teller in the Theoretical Group, looking into the possibility of a weapon based not on nuclear fission but on fusion. From February 1944 until February 1946, Mary Argo performed calculations on deuterium-deuterium and deuterium-tritium burning. After the war, the couple returned to Chicago with Teller and started a family while Harold completed his Ph.D.

Back in Chicago, the Argos lived on the top floor of Maria Mayer's large house in Hyde Park. They have pleasant memories of Mayer's motherly warmth and lack of affectation. In 1948 they returned to Los Alamos, where Mary worked until 1960 doing opacity calculations and theoretical studies of how materials change at different temperatures and pressures. She bore and raised four children while working, and continued to enjoy both her work and the respect of her colleagues.

Mary Argo was the only female staff member officially invited to see the test of the first nuclear weapon. She and her husband drove with camping gear to Chipadera Peak near Socorro, New Mexico, about thirty miles from ground zero. Through binoculars, they could see the tower holding the "gadget." Just at sunup on July 16, 1945, they saw the world's first nuclear explosion light up the New Mexico desert (Argo 1991; Teller 1991).

Although Mary Argo was the only female staff member officially invited, quite a few women who had not been invited saw the Trinity test anyway. One of them was Joan Hinton, a young graduate student in physics working at Los Alamos, who sneaked into the area on the back of a motor scooter owned by a fellow worker.

In many ways, Hinton typified the talent and exuberance of the young physicists who had been pulled from their graduate studies to assist in the war effort. Male or female, they possessed considerable intelligence, scientific training, and physical energy, and in many cases a notable dislike of regimentation. Their high spirits and the creative activities they invented to relieve the boredom of off-duty hours in isolated Los Alamos account for a generous portion of the mystique of the secret laboratory and certainly help to explain how its character differed from that of Chicago, Hanford, and Clinton.

Hinton's high-school chemistry teacher introduced her to the tracks visible in a simple cloud chamber. She became fascinated with atomic physics in general and cloud chambers in particular. While majoring in physics at Bennington College in Vermont, she constructed her own cloud chamber for her sophomore project, learning to operate a lathe, construct control circuits, and make patterns at a nearby foundry.

Bennington offered winter field periods to its students, and Hinton spent her final two working at Cornell University, where her brother was studying dairy farming. When the cyclotron next door to the lab in which she worked broke down, she happily helped to repair it, assisting with its vacuum problems, constructing gauges, and soldering things back together. In the process, she got to know some of the physicists who would later be her coworkers on the Manhattan Project.

When Hinton graduated from Bennington, she began graduate study at the University of Wisconsin. (Cornell refused to admit a woman to its graduate program in physics.) Bennington had not offered calculus, so Hinton studied it by correspondence; she also took a correspondence course on the kinetic theory of gases. She acquired a reputation as a daring skier on the long slopes of the Wisconsin campus.

World War II had begun by this time, and the pressures of the war-related research projects were being felt in university laboratories. Hinton was beginning construction of yet another cloud chamber when the Van de Graaff accelerator next door disappeared, along with the physicists who ran it. When she was recruited by the physicists from the cyclotron crew at Cornell, who were now at Los Alamos, the University of Wisconsin cooperated by making the examination for her master's degree a very informal one: She sat on the floor of the lab, answered a few questions, and showed the committee a photograph of her cloud chamber (J. Hinton 1990).

Stanislaw Ulam, a mathematician on the Wisconsin faculty who subsequently worked on the Manhattan Project, recalls Hinton's sudden departure:

> Joan was taking a course I gave in classical mechanics. One day she appeared in my office in North Hall to ask if I could give her an examination three or four weeks before the end of the term so that she could start some war work. She produced a letter from Professor Ingraham, the chairman, authorizing me to do that. She was a good student, a rather eccentric girl, blonde, sturdy, good looking. Her uncle was G. I. Taylor, the English physicist. She was also a great-granddaughter of George Boole, the famous nineteenth-century logician. I wrote a number of questions on the back of an envelope; Joan took some sheets of paper, sat down on the floor with her notebook, wrote out her exam, passed, and disappeared from Madison. (Ulam 1996, p. 144)

At Los Alamos, where she arrived in February or March of 1944, Hinton was assigned to work with Enrico Fermi. At this time, Fermi had two working groups: One was mainly theoretical, though it also performed assorted experiments; the other was located in a canyon some distance from the main building, where it was building the first reactor to use enriched uranium for fuel. Hinton joined the second group, which actually built two reactors to test assemblies of enriched uranium and later plutonium. The first, which achieved its first chain reaction in May 1944, ran at low power to perform preliminary tests and identify unexpected problems. The group then cannibalized the first reactor to construct a much more powerful one as a strong neutron source for experimental work. The group called the reactor "the water boiler"—a nickname that referred to the water used to cool it.

The canyon group worked in relative isolation to protect the rest of the laboratory from reactor accidents. To reach their "lab," which consisted of a shop and a few offices, the scientists had to climb or drive down a cliff from the main laboratory. The group built its reactor from scratch. Hinton piled beryllium blocks around the spherical core and constructed electronic circuits for counters. She helped to design and construct the control rods for the second version of the system. The first reactor used uranyl sulfate solution, called "soup," as fuel. In the second version, the soup was replaced by uranyl nitrate, which allowed easier extraction of plutonium. Gaseous fission products were vented through plastic tubing that ran up a nearby pine tree.

Hinton described the first test of the low-power ("lopo") reactor:

> When it came time to test the "lopo" there was great excitement in our little group. Just for fun we put a chart up on the wall (actually a graph)—plotting amount of enriched uranium against activity of the counters when the control rods were taken out. The point was to see just how much U235 it would take to get a self sustaining reaction. And to test the theoreticians who were attempting to calculate this figure from the configuration of "lopo". In the spirit of the test, many of the "big shots" came down and signed their initials on the graph at the place they expected it to "go critical". Fermi, Bethe, Christy, Kerst,—all put their initials on the chart—and of course all of us did too (though we had no idea where it would go critical having made no calculations—pure guessing on our part—).

Hinton also worked on the high-powered ("hypo") reactor.

> As we started building the "Hypo", the first small sphere of plutonium arrived. (I always thought it was the core of the Nagasaki bomb—but Phil Morrison told me in 1982 when I saw him that he didn't think it was... I remember quite distinctly that this was the first plutonium sphere to come out of the Washington plant).
>
> I remember so well when this first ball of plutonium arrived at Omega. There was a quiet stir through our "hypo" group. As I came to work one day some one told me in a low voice "the plutonium* has come. It's in that little room. It feels warm to the touch". I went in and had a peek—no one else was in that room at the moment. I didn't dare actually touch it without anyone around to be sure I was doing it right. It was a round metal ball about as big as a baseball. I looked at it for some time, then came away... (*Also in those days we did not call it plutonium—but rather its code name "49.") (J. Hinton 1990)

The Los Alamos scientists took plenty of risks, but this user-friendly handling of plutonium wasn't really one of them. In solid form, plutonium isn't particularly dangerous. You can handle it with paper gloves, or even eat it without serious consequences. The work on the water boiler was far more dangerous, both because of the properties of the reactor itself (high radioactivity, the chance of meltdown) and because the crew knew relatively little about the fissile materials they were handling. On one occasion, at about the time of the Trinity test, Hinton was analyzing some films that had been irradiated in the reactor when she heard shouting. She and her companion ran outside to find the workers from the reactor room crouching behind a mound of dirt piled up by a bulldozer. Inside, radiation alarms were sounding. Her companion ran to get a radiation counter, and they joined the others. To everybody's relief, the counter showed little radiation present. When they reentered the lab, they found that one of their colleagues had had an accident with the first sample of plutonium he had tried to immerse in water. As water covered the sample, slowing neutrons and reflecting them back into the plutonium, the sample went critical. The water began to glow blue with Cherenkov radiation. Fortunately, the physicist kept his head and dropped the water level, stopping the chain reaction. He lost his hair from radiation exposure, but otherwise escaped unharmed.

On another occasion after the war, a very serious accident occurred, which Hinton described:

> One day I came down to work early after lunch. I was just turning the car around to park it ready to drive out, when Bill Starner came out of the building. He said, "Harry's had an accident, could you drive him to the hospital?" Bill Starner, Harry Daghlian and I were the only ones there. Harry sat next to me in the front seat on the way up. He kept rubbing his right hand on his knee. The hand looked a bit swollen. He said he had come down early after lunch. In the morning they had been placing paraffin blocks in different positions. Since none of the others had come yet, he started "playing" with it on his own. He picked up a piece of paraffin thinking to place it on the opposite side of the [plutonium] sphere, but as he carried [it] over the sphere it slipped out of his hand and landed on the sphere. The sphere went blue. He knocked the paraffin off of the sphere with this right hand. Now the hand felt very strange. He was worried that he might lose his hand. When we got to the hospital up on the site, the doctors took all Harry's metal objects out of his pockets—keys, knife, etc., and gave them to me. They told me to get a "count". I drove right back down to Omega and put them in our geiger counters. But they

were so "hot" the counters just jammed. For many hours there was no way to get a count. Finally we were able to get a curve of the radioactivity.

I never saw Harry again. Bob Carter and Harold Hammel went to see him almost every day—and every day they reported to the rest of us Harry's condition. He was the first case of radiation sickness any of the doctors had ever seen or handled. He very gradually just disintegrated. All the cells in his body had been damaged way down inside. There was nothing the doctors could do. They packed him in ice. Finally his hair began to fall out and his mind began to be affected. One of the last things I remember Bob Carter telling us was that Harry was worried about his draft card. Over and over he kept asking Hammel and Carter to find his draft card. He would concentrate on just this one thing and nothing else.

It took Harry a month to die. It was a very sobering experience for everyone on the site. And it added fire to the immediate reaction of the Los Alamos scientists against using the bombs on Hiroshima and Nagasaki. When 150,000 Japanese disappeared into a huge lethal mushroom of radiation, Harry's death helped bring the lesson home. (J. Hinton 1990)

Hinton was part of the group of graduate students and young scientists who surrounded the Italian physicist Enrico Fermi. Possibly the most gifted scientist of his generation, Fermi both performed theoretical calculations and conducted experiments. He liked bright young people and enjoyed teaching them physics by discussing his ideas with them. At one point, Hinton says, Fermi suggested that they collaborate on a paper. He changed every sentence, but still gave her coauthorship.

Everyone worked very hard, but they played just as hard, taking daylong hikes and wild skiing adventures, in which Fermi, who was then in his early forties, was the ringleader. The shop in the canyon could be used to make steel edges for skis as well as reactor components. Hinton's intelligence and physical daring well qualified her for this group. She also played violin in a quartet that included Edward Teller and Otto Frisch.

As the time to test the atomic bomb grew closer, tension mounted at Los Alamos. According to Hinton, security was tight, and information was given out on a need-to-know basis. The graduate students doing basic experiments with the water boiler were not formally involved in this first nuclear test, but they shared laboratory space with the group that had produced the plutonium for the test, so Hinton and her friends knew very well what was happening—including the time and place of the detonation. They decided to watch the test from a low hill twenty-five miles from ground zero at Alamogordo.

Dodging Army patrols, Hinton and a friend rode to the mound at sundown on the friend's motorcycle. They hid the cycle, then waited all night; the detonation, which had been planned for midnight, was delayed by thunderstorms. Just before dawn, they saw an amazing bath of light and felt heat as the bomb went off:

> It was like being at the bottom of an ocean of light. We were bathed in it from all directions. The light withdrew into the bomb as if the bomb sucked it up. Then it turned purple and blue and went up and up and up. We were still talking in whispers when the cloud reached the level where it was struck by the rising sunlight so it cleared out the natural clouds. We saw a cloud that was dark and red at the bottom and daylight on the top. Then suddenly the sound reached us. It was very sharp and rumbled and all the mountains were rumbling with it. We suddenly started talking out loud and felt exposed to the whole world. (J. Hinton 1990)

To their surprise, daylight showed Hinton and her friends that their secret viewing area also held ten or fifteen other uninvited spectators (J. Hinton 1990). Great minds, apparently, think alike.

Margaret Ramsey Keck came to Los Alamos after receiving her undergraduate degree in physics from the University of Rochester, where she had worked with Robert Marshak doing calculations in astrophysics. Marshak, a young nuclear physicist who would later become a deputy group leader at Los Alamos, suggested that the Manhattan Project hire her, and she joined the project as a junior scientist in the spring of 1945. Because she wanted a change from theoretical work, she was assigned to a group working on the implosion explosives—the explosives that would be used to compress the sphere of plutonium into a supercritical mass. For safety, the Explosives Group was stationed out of town in an area called South Mesa.

The explosives used to compress the fissile mass had to be very carefully screened, because nonuniform grain sizes could cause variations in the timing of the explosion, and any distortion in the shock wave could cause an incomplete detonation of the bomb. Ramsey took photomicrographs—photos of magnified images—of the explosive used in the detonators (Keck 1991).

Ramsey worked on the project with a young SED from Brooklyn, New York, named Benjamin Bederson, who had two years of college physics and a background in the Army Specialized Training Program, which taught soldiers to be engineers. Bederson had been recommended for assignment to the Manhattan Project by a well-meaning commanding officer who assumed that he would be sent closer to his home. Instead, he was sent to Los Alamos, and eventually to Tinian to help assemble the weapons used in combat. Bederson recalls Ramsey's red hair and sense of humor (Bederson 1996).

After the war, Ramsey worked on beta-decay experiments until 1946, when she left Los Alamos to study at Indiana University. She married James Keck, a physicist, and moved to Cornell University, where she completed course work for her master's from Indiana while working as a physics teaching assistant and then as a research assistant. At Cornell, she studied the tracks of nuclear events in photographic emulsions. She and her husband moved to the California Institute of Technology after several years, where Ramsey found similar work. She left physics soon after arriving in California, however (Keck 1991).

One other female physicist, Jane Roberg, remains something of a mystery. Roberg obtained her Ph.D. from Duke University in 1942 and moved to the University of California as a research fellow (*American Men and Women of Science*). Like most physicists, she was recruited for the Manhattan Project. At Los Alamos, she worked on the calculations for the fusion weapon. Very much a loner, she is usually described as attractive and somewhat mysterious. We have been unable to discover any details about her departure from Los Alamos or her subsequent work.

## In Conclusion

We've described the work of female physicists at the major sites of the Manhattan Project. However, the project sent its tentacles into a variety of industries, and these industries had their share of female scientists as well. In 1943, for example, Helen Jupnik accepted a job at the American Optical Company, though she almost immediately obtained a leave of absence to return to Princeton University. Jupnik received her Ph.D. in optics from the University of Wisconsin, was appointed a Huff Fellow at Bryn Mawr College for the following academic year, and received a fellowship to study in Berlin in 1941–42. She then worked at Princeton and for the National Research Council.

At Princeton, Jupnik worked on experiments measuring the resonance absorption of neutrons in uranium. These basic nuclear studies attempted to determine how a neutron's energy affected its chance of producing a fission reaction in another nucleus. The data were collected by Geiger counters and reduced with electronic aids. Jupnik's work fell very much in the mainstream of the physics done by the Manhattan Project, and some members of her group transferred to the Met Lab in Chicago. Jupnik blames sexual discrimination for the fact that she did not play a larger role in the Manhattan Project (*American Men and Women of Science*; Creutz 1991; Kathren et al. 1994).

One female physicist obtained an appointment to the committees guiding the scientific work of the project after the war. Gertrud Nordheim, a German-educated theoretical physicist, had emigrated from Germany to the United States in the early 1930s. She and her husband, Lothar Nordheim, who had been her professor at Göttingen, came to Oak Ridge from Duke University "just long enough to give birth to their son Ricky," as Lothar put it, then returned to the Duke faculty in 1947. Their German extraction apparently did not pose a problem for security. Lothar Nordheim described their first visit to Oak Ridge:

> We drove . . . to Oak Ridge for a first glimpse of our new home for the next few years. We got our due initiation to military security when the guards found a half-empty bottle of whiskey among our belongings and promptly confiscated it. We were not alert enough to demand that its contents be poured out immediately. Inside the reservation we were duly impressed by the enormous amount of construction going on. (Nordheim 1971)

Gertrud worked at Oak Ridge on calculations of neutron diffusion in the atomic piles. Her husband described one of her experiences en route to work:

> The long working days were not made any shorter by the bumpy rides to the distant lab. But they were not without their amusing incidents. One day on the bus, my wife, then visibly pregnant, engaged in a spirited discussion with Charles Coryell. She mentioned the element rhenium, and the guard thought she had said "uranium." Due to her delicate condition she got away with a mild reprimand, but not until after she had been given a lengthy exhortation in my presence.

In the summer of 1946, both Gertrud and Lothar were appointed consultants in physics to Los Alamos (Truslow and Smith 1961). Despite her appointment to this high-level group of theoretical leaders, her name never appeared in the subsequent work of the laboratory or in physics. She was killed in a traffic accident while visiting her family in Germany in 1947.

Although relatively few female physicists participated in the Manhattan Project, they contributed to almost all phases of the work. Women were represented among the junior physicists proportionally to their participation in the profession, but no women sat among the small group of the project's senior leaders. There are at least two factors that may have contributed to their absence.

First, World War II followed on the heels of the Great Depression. Academic jobs were at a premium, and there are many reports of disgruntled young faculty members protesting the granting of scarce jobs to distinguished European immigrants. Except at women's colleges, faculty positions were almost never offered to women. A woman wishing to pursue research in physics could either accept a junior position at a research university, as Maria Mayer did, or take a faculty position at a women's college, none of which was considered a leader in research. In the United States, therefore, there were no senior female professors or university administrators automatically qualified to assume leadership roles in the Manhattan Project. Gladys Anslow, who as a section chief of OSRD's Office of Field Service was as close as any woman was to being a member of the Manhattan Project leadership, obtained her post on the basis of her experience leading the Physics Department at Smith, a women's college.

Physicists' view of the women in their ranks is shown by a 1936 letter from Robert Millikan to President Few of Duke University, who had asked Millikan his opinion of hiring a female physicist. Millikan wrote, in part:

> Women have done altogether outstanding work and are now in the front rank of scientists in the fields of biology and somewhat in the fields of chemistry and even astronomy, but we have developed in this country as yet no outstanding women physicists. In Europe Fraulein Meitner of Berlin and Madam Curie of Paris are in the front rank of the world's recognized physicists. I should, therefore, expect to go farther in influence and get more for my expenditure if in introducing young blood into a department of physics I picked one or two of the most outstanding younger men, rather than if I filled one of my openings with a woman. I might change this opinion if I knew of other women who had the accom-

plishments and attained to the eminence of Fraulein Meitner. I know of no other living woman who has had anything like her accomplishment or has prospects in the future of having such accomplishment.

Also in the internal workings of a department of physics at a great university I should expect the more brilliant and able young men to be drawn into the graduate department by the character of the men on the staff, rather than the character of the women.

These considerations relate more to the graduate work than to the undergraduate. In a coeducational institution where there are many women students it is undoubtedly also desirable to have for pedagogical purposes women instructors, but only in very exceptional cases would I think that the advance of graduate work would be as well promoted by a woman as by a man. (Rossiter 1982, plate 25, pp. 192–93)

Another reason for the lack of women among the leaders of the Manhattan Project is that—as Millikan mentioned in his letter—the genesis of nuclear physics had taken place in Europe. It was therefore hardly surprising that many of the leaders in the field were European. During World War II, many of the male leaders in nuclear research, such as Enrico Fermi and Niels Bohr, emigrated to the United States and played leading roles in the Manhattan Project. The female leaders, such as Irène Joliot-Curie and Lise Meitner, remained in Europe, where the war greatly slowed their work.

Even senior women, such as Elda Anderson, worked as staff scientists and junior scientists rather than group leaders and members of guiding committees. From these relatively junior positions, nonetheless, they made significant contributions to the physics of the Manhattan Project.

Marie and Pierre Curie in their Paris laboratory. (*Argonne National Laboratory, OLDNEG #1-1094.*)

Irène and Frédérick Joliot-Curie working on the experiments in radiochemistry that earned them a Nobel Prize in 1934. (*French Embassy—Information Division, courtesy AIP Emilio Segrè Visual Archives.*)

Lise Meitner shown with Otto Hahn in her laboratory outside Berlin before she was forced to flee to Sweden. *(AIP Emilio Segrè Visual Archives)*

Maria Mayer talks with Enrico Fermi. *(AIP Emilio Segrè Visual Archives, Goudsmit Collection)*

The Los Alamos woods form a backdrop for a discussion among Joan Hinton, Robert Carter, and Enrico Fermi. *(Argonne National Laboratory, OLDNEG #1-1306)*

Formal security photographs of women who worked at Los Alamos show how very young most of them were.
*a.* Mary F. Argo.
*b.* Naomi M. Livesay.
*c.* Elda Anderson.
*d.* Elizabeth R. Graves.
*e.* Jane Robert.
*(Los Alamos National Laboratory Photographs, Negs. #78-9657, 78-9680, 78-9656, 78-9670, 78-9692, respectively)*

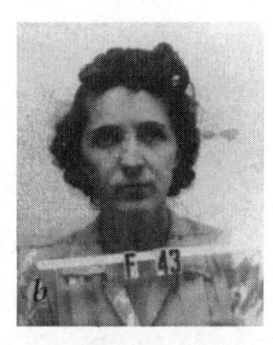

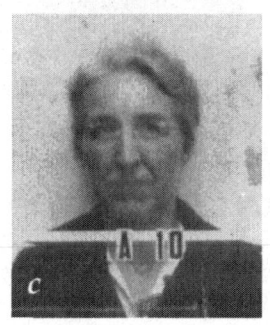

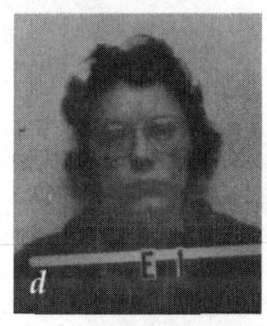

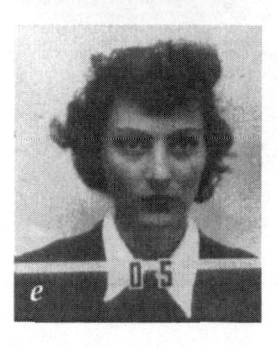

Leona Marshall started work on the Manhattan Project in her twenties. *(Argonne National Laboratory, NEG #1-790)*

Chien Shiung Wu wearing traditional Chinese dress climbs a ladder to fine tune her experiment. *(AIP Emilio Segrè Visual Archives)*

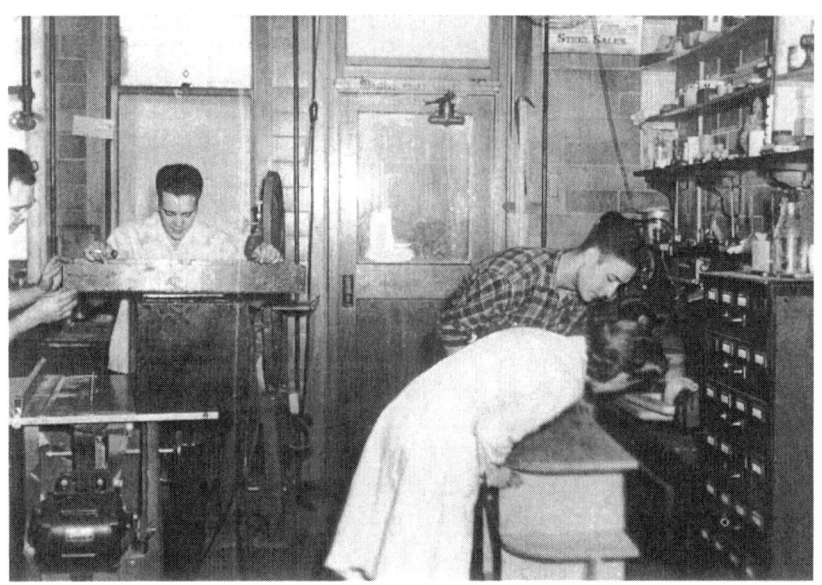

Kay Garvey works with Reid Harding, Emil Fuian, and Bob Young in their laboratory, Room 12, at the Met Lab in Chicago in spring 1944. *(Courtesy of the Lawrence Berkeley National Laboratory)*

Eleanor Hauk works as a draftswoman at the Clinton Laboratory under the direction of health physicist William Ray. (*Monsanto Magazine*, February 1946 issue)

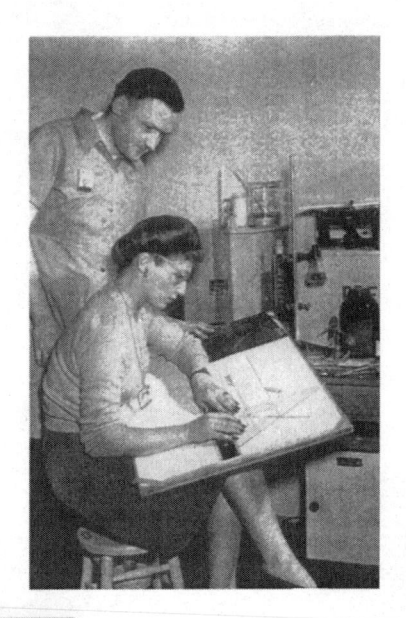

General Groves reviews reports with his secretary Jean O'Leary, who functioned as his second in command although she held no title. *(Department of Energy photograph by James E. Westcott, Neg. #2044)*

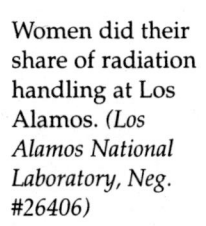

Women did their share of radiation handling at Los Alamos. *(Los Alamos National Laboratory, Neg. #26406)*

Women participated in early studies of the biological effects of radiation, as in this photograph from the Oak Ridge Hospital laboratory. *(National Archives, RG 434-OR, Box 12, Notebook 37, 1310-10)*

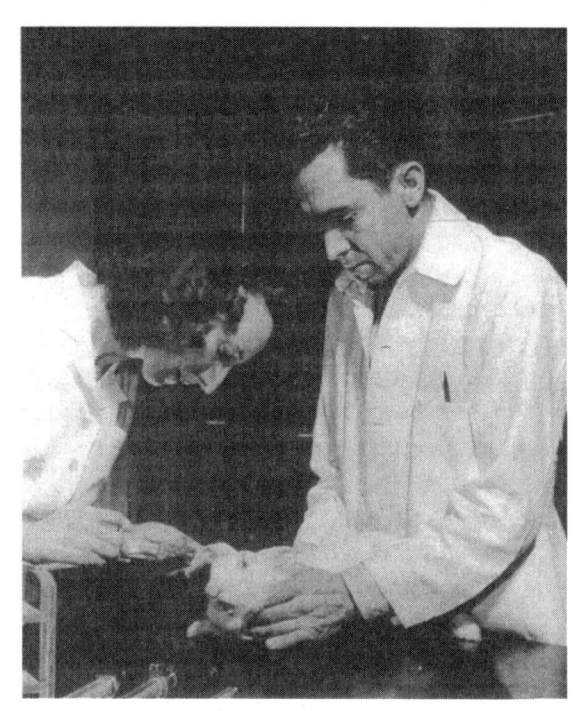

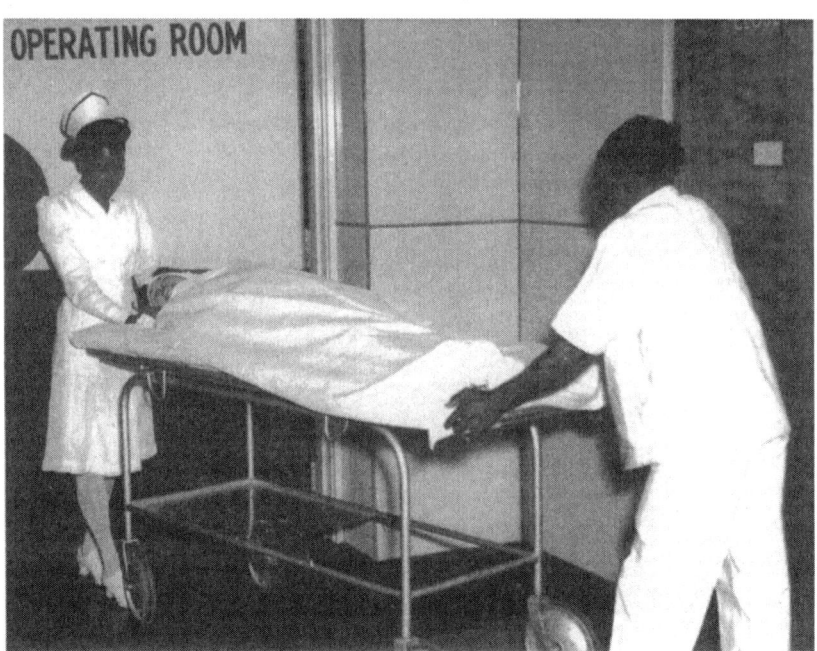

Nurses at the hospital in Oak Ridge. *(National Archives, RG 434-OR, Box 12, Notebook 37, 1310-10)*

# 4  The Chemists

WITHOUT CHEMISTS, the bombs designed by the physicists could never have been built. Manhattan Project chemists faced unprecedented challenges in working with newly discovered radioactive elements. They gradually charted the complex chemistry of uranium, a frontier field in its own right, and they produced the scores of new materials required for the manufacture of fissionable uranium-235, plutonium, and the bomb itself.

For example, Helen Blair Barlett took temporary leave from her job at the A.C. Spark Plug Division of General Motors to study ceramics for the Manhattan Project at MIT, where she developed a nonporous porcelain for the interior construction of the bomb (Rossiter 1995). Barlett, a mineralogist who had received her Ph.D. from Ohio State in 1931, had applied her training in mineralogy and petrology to the study of alumina ceramics, especially concerning their use as insulation for spark plugs. Her work earned her several patents and considerable status at General Motors, so she came to the war effort as a senior scientist (Kass-Simon et al. 1990).

The work of the Manhattan Project's chemists, like that of the physicists, carried its dangers. They were used to working with microgram quantities of irreplaceable materials, but no one had ever tried to handle large quantities of radioactive matter. The biological effects of plutonium and uranium were not even known when the Manhattan Project began. Time pressure forced the project to take chemical processes directly from the lab bench to industrial-scale production, skipping the normal pilot-plant stage.

Many female chemists worked on the project, but we know little about some of them. Female chemists, like their male counterparts, frequently worked outside their areas of expertise as the needs of the project changed. Some women changed fields entirely. Many married during the war and changed their names.

Like other women in war work, many female chemists had to cope with discrimination and, in some cases, lack of proper training and dangerous working conditions. They frequently lived in spartan quarters far from home and family, and they were sometimes homesick. Still they made valuable, creative contributions to the war effort (Wise and Wise 1994; Rossiter 1995).

## Early Work on the East Coast

Uranium had limited industrial applications before its military implications were realized (though it *was* used as a glaze for pottery), so few chemists had examined the compounds it formed. In 1939, they didn't even know the melting point of uranium metal. The metal was known to be highly pyrophoric (it burned when exposed to air), toxic, and extremely difficult to handle. With the discovery of fission, chemists faced their first challenge: producing pure uranium metal.

The production effort began at Iowa State College under the direction of Frank Spedding. Spedding and his team learned to process uranium ore and to produce chemically pure uranium metal. Elsie Fornafelt worked in the analytical laboratory at Iowa State College (Svec 1995), and other women were probably involved as well, though we don't know their names.

Once the Manhattan Project had procured pure uranium metal to fuel the early atomic piles, project scientists recognized that construction of a bomb required several kilograms of uranium-235, the isotope that fissioned with thermal neutrons. If the core of the bomb were not substantially pure uranium-235, the fission reaction would blow the bomb apart before most of the nuclear energy could be released. Natural uranium consists almost entirely of the isotope uranium-238; uranium-235 accounts for only 0.7 percent. The two isotopes are chemically identical, so separation processes must depend on the tiny difference in mass between them. No one had ever attempted to separate isotopes of any element, except hydrogen, on an industrial scale.

Harold C. Urey, a chemistry professor at Columbia University, decided to separate the uranium isotopes by a process known as gaseous diffusion. In this process, a gas of natural uranium is pumped through thousands of barriers with tiny holes in them. The lighter isotope, uranium-235, gets through the holes slightly faster than the heavier uranium-238, so the gas collected at the end of the process contains more than 0.7 percent of uranium-235. This gas was called "enriched," though at least 20 percent uranium-235 is needed for an explosion. (In modern weapons the percentage is over 90.)

Unfortunately, the only chemical form of uranium that is a gas at room temperature is uranium hexafluoride, a highly corrosive substance that tended to destroy the barriers through which it passed. Thus, in addition to figuring out how to produce the hexafluoride gas and how to recover uranium as a metal after the separation, it fell to the chemists to find a barrier material that wouldn't corrode.

The Columbia laboratory desperately needed skilled chemists. Mary Holiat Newman received a bachelor's degree in chemistry from Barnard College and intended to go to graduate school, but needed to earn some money first. The college employment office directed her to the Columbia branch of the Manhattan Project, which eagerly hired her. She worked on the development of components for the gaseous-diffusion plant (Howes and Herzenberg 1993). Like many other project workers, she married a fellow employee. At the end of the war she accompanied her husband to Oak Ridge, where she later became chief of the Evaluation Section of the Technical Information Division.

Lotti Grieff was teaching chemistry in New York when she was recruited by the Columbia lab. She had received an M.S. from Cornell University and a Ph.D. in chemistry from Columbia University in 1931 (*American Men and Women of Science*), and she had worked as an analytical chemist for the U.S. Rubber Company for four years. A chemist with this type of experience was a real find. She joined the gaseous-diffusion project in 1942 (Rossiter 1995). After the war she returned to teaching, enlivening her lectures with stories of her experiences in the Manhattan Project (Schriesheim 1992).

Susan Chandler Herrick was not a Columbia graduate, but she had studied physical chemistry at Sarah Lawrence with the future Nobel laureate Maria Mayer—in fact, Herrick was the only student in the class. Mayer easily got her a job with the Manhattan Project. Herrick had completed all but one semester of college and had been accepted to medical school on the condition that she not marry. She found herself with a year to fill between college and medical school (as it turned out, she never got to medical school, deciding to marry after all), and in 1943 she accepted a job with Harold Urey, who by this time couldn't sleep for worrying that the Germans would figure out isotope separation first. He was desperate for talented, trained chemists.

Herrick worked in Mayer's group on problems in uranium chemistry, which included synthesizing and crystallizing compounds of uranium that might be of interest to the isotope-separation project. She developed techniques to produce single crystals of uranium-235 from samples of a few hundred milligrams, and she contributed to the development of improved nickel barriers for the gaseous-diffusion plants (Herzenberg and Howes 1993). Herrick claims that she knew little about the chemistry she did, but that her knowledge of German came in very handy, as most of the prewar work on uranium compounds had been published in that language.

Radioactivity wasn't the only hazard of the lab. While constructing plastic covers for solutions of uranium, Herrick managed to catch her hair in a drill press and lose a couple of square inches one week before her wedding (Herrick 1991).

Lotti Grieff and Mary Newman were recruited to the project because they held degrees in chemistry from Columbia. Other women simply happened to apply for jobs at the right moment. For example, Eleanor Reace joined an outpost of the Manhattan Project at MIT when the Navy assigned her husband to the radar project and she needed a job in the Boston area. She handled and tested alloys of uranium-235, one of the numerous steps on the way to deciding which material could best be used in the core of a fission weapon.

## MEANWHILE, ON THE WEST COAST

As East Coast universities struggled to understand uranium, chemists at West Coast schools studied the newly discovered radioactive element plutonium. In May 1940, the physicists Edwin M. McMillan and Philip H. Abelson bombarded uranium in the sixty-inch cyclotron at the University of California at Berkeley and found that they had produced a new element—the ninety-third in the periodic table—which they eventually named neptunium. Then, using the same cyclotron, Glenn Seaborg, Edwin McMillan, Joseph Kennedy, and Arthur Wahl discovered the ninety-fourth element, plutonium, by deuteron bombardment of the uranium isotope U-238. By May 1941, the Berkeley group had established that Pu-239 was 1.7 times as likely as U-235 to fission under the conditions that would occur in a bomb.

Project scientists were delighted by this news, which meant that if the uranium isotopes could not be separated on a large scale, a bomb could be constructed of plutonium instead. The use of plutonium raised huge challenges of its own, however. The necessary quantities of the element would have to be produced in the fuel of an atomic pile that sustained a chain reaction. The highly radioactive fuel rods would be removed from the pile, cut in pieces, dissolved, and treated to separate the plutonium from the uranium and other elements. Developing a process that could work on a large scale with the lethally radioactive fuel rods, let alone understanding the chemical and radiation properties of microscopic samples of the new element, posed enormous problems.

Most of the female chemists on the West Coast worked at the University of California at Berkeley. They worked in makeshift quarters, borrowing equipment where they could find it. Marjorie Woodard Evans, a graduate student at Berkeley, worked as a research assistant from 1942–45. Her research group studied the plutonium extracted from fuel rods and how it might be recovered from the solutions that contained it. She received her Ph.D. in physical chemistry from Berkeley in 1945, and went on to work at various other institutions, including the Stanford Research Institute, where she became Executive Director of the Physical Science Division (*American Men and Women of Science*; Kathren et al. 1994; Seaborg 1977).

Another Berkeley chemist, Margaret Melhase, worked with Seaborg in 1940–41, while she was still an undergraduate, and helped prepare the original chart of the isotopes. As one of Seaborg's undergraduate research students, she was a codiscoverer of the important isotope cesium-137. Seaborg describes her undergraduate research project:

> In the late fall of 1940 I asked an undergraduate student, Margaret Melhase (now Mrs. Robert Fuchs), to take some uranium which had been bombarded with neutrons furnished by the 60-inch cyclotron and make chemical separations designed to look for hitherto unknown radioactive fission products such as cesium. She performed her chemical separations on the top floor of the old "Rat House" which even at that time was an ancient ramshackle wooden building dating from the earliest days of the Department of Chemistry at Berkeley. Her measurements of the radioactive decay and radiation absorption properties were performed through the use of a Lauritsen quartz fiber electroscope situated in the cavernous auditorium of abandoned East Hall. . . .

Miss Melhase also used a Lauritsen electroscope situated on the third floor of Gilman Hall in the latter stages of her research after she had graduated to this higher status. She continued her work until the summer of 1941, by which time she had established the presence of a very long-lived radioactive fission product in the cesium fraction which on the basis of subsequent work we can now identify as being due to the 30-year cesium-137. (Seaborg 1970)

Melhase received a B.S. in chemistry from Berkeley in 1941. She went to work for the Philadelphia Quartz Company in El Cerrito, California, but rejoined the Manhattan Project at Berkeley in 1944 and continued work until 1946 (Kathren et al. 1994). In 1945, she married Robert Fuchs—another Manhattan Project romance—and stopped doing technical work.

There were many women at Berkeley about whom we know very little. Virginia S. Holmes, for example, was born in Mount Vernon, Iowa, in 1919, and attended Oberlin College from 1936 to 1939. Her title was lab assistant, but she served as secretary to Wendell Latimer, who led Berkeley's plutonium research, from 1942 to 1943 (Seaborg 1977; Kathren et al. 1994). Peter Yankovitch tells of several women who actually trained male graduate students to run the chemical labs that analyzed the purity of uranium and plutonium samples, though he doesn't remember these women's names. Some of the women had extensive experience in forensic analytical chemistry and were of great assistance to their young male bosses.

## FEMALE CHEMISTS AT THE METALLURGICAL LABORATORY

The Metallurgical Laboratory was established on the campus of the University of Chicago in early December 1941 to consolidate the research projects scattered around the country. Its primary goals were to develop chain-reacting nuclear piles, to devise methods of extracting plutonium from the irradiated uranium, and to start weapons design. Arthur Compton, the project leader, brought scientists already engaged in small-scale studies to this central location. In April 1942, much of the pioneering work on plutonium was transferred from California to Chicago. By consolidating his personnel, Compton concentrated the existing expertise on plutonium and fission piles, although the Columbia group continued its studies of gaseous diffusion. The consolidation also eased security problems.

By 1943, the Met Lab comprised divisions dealing with nuclear physics, chemistry, engineering, metallurgy, and the newly troublesome health effects of plutonium and uranium. The Chemistry Division itself was divided into sections dealing with the chemistry of final products (uranium and plutonium), radiation chemistry, the chemistry of by-products (other isotopes produced in the piles), and analytical chemistry. As the work expanded and more people arrived, additional groups were formed. The Heavy Isotopes Division worked with enriched samples of uranium and plutonium. Control Analysis monitored the purity of uranium ore. Solvent Extraction developed a process to get plutonium out of fuel elements and into solution, and the Recovery Group worked on ways to extract it from solution and into metal form (Kathren et al. 1994).

The Met Lab was spread over a number of buildings on the campus of the University of Chicago, including the west stands of Stagg Field, where there was space to build atomic piles. Administration was lodged in the Museum of Science and Industry, off Jackson Park near Lake Michigan. Metallurgy, biology, and an additional machine shop were in a former bottling plant code-named Site B and referred to as "the brewery" (Greenbaum 1971). In 1943, the reactor experiments moved off-campus from Stagg Field to Site A, in Argonne Forest, eighteen miles southwest of the university in the Cook County Forest Preserves. Argonne National Laboratory now operates a few miles west of the original location of Site A.

In 1942, even the words for fissile elements were classified. Plutonium was referred to as "product" and uranium as "tuballoy" (Friedell). Work was tightly compartmentalized, again for secrecy. Only micrograms of plutonium had been produced in cyclotron bombardments at Berkeley, so the Met Lab chemists had to work with quantities of plutonium too small to see. Finally, on August 20, 1942, the Met Lab precipitated a visible particle of pure plutonium (Graf 1994).

The first weighing of a sample of plutonium took place at the Met Lab on September 10, 1942. The sample of the dioxide of plutonium ($PuO_2$) weighed 2.77 micrograms, and measuring it required a balance made of quartz fibers so thin that they could barely be seen (Seaborg 1989). Photographs of the work leading up to this measurement show several women. In one, an unidentified, dark-complexioned young woman wearing a laboratory coat is engaged in producing the quartz fibers for the balance. Another photograph shows Virginia Hawley Bisberg, wearing a picture ID badge, adjusting the completed balance.

The chemistry of plutonium turned out to be complicated. Not only were there different isotopes, but pure plutonium existed in six forms, differing in crystal structure and density (allotropes). The element also exhibited four different oxidation states. The group struggled to remove light-element impurities down to less than a part per million (Seaborg 1989).

Isabella Lugoski Karle became the most distinguished female chemist employed at the Met Lab. The daughter of Polish immigrants, she was born in Detroit in 1921. Although she could not speak English when she entered first grade, she won a four-year undergraduate scholarship to the University of Michigan. Denied a chemistry teaching assistantship because she was a woman, she continued her graduate work with the help of a fellowship from the American Association for University Women. She married Jerome Karle, a fellow graduate student in crystallography, in 1942. Jerome Karle subsequently became a renowned crystallographer and shared the 1985 Nobel Prize in Chemistry (*American Men and Women of Science*; Julian 1986; Kathren et al. 1994).

Isabella Karle completed what she describes as a "fast" Ph.D. in experimental physical chemistry in 1943, on her twenty-second birthday. Her husband finished his Ph.D. in theoretical physical chemistry in the same year. After graduation, the couple moved to Chicago to work on the Manhattan Project (Julian 1990; Kathren et al. 1994; Bailey 1994). When Isabella started work at the Met Lab on January 1, 1944, Glenn Seaborg wrote in his journals: "Jerome Karle's wife Isabella started work in Section C-I today. L. O. Brockway of the University of Michigan wrote on November 19 that she and her husband had made the best records there of any recent graduates in chemistry. Those who know Mrs. Karle praise her as 'brilliant' and having a 'most pleasing personality'" (Kathren et al. 1994, p. 365).

Isabella Karle studied the chemistry of transuranic elements and the synthesis of plutonium compounds (Howes and Herzenberg 1993; Roscher 1993; McMurry 1995). She had an appointment as an associate chemist, and she was responsible for vapor phase reaction of carbon tetrachloride with plutonium dioxide (Kathren et al. 1994). Using the best existing microtechnology, she grew crystals of plutonium chloride for the first time. She repeated the experiments growing crystals in a number of different ways and demonstrated the stability of the new compound (Karle 1991; Howes and Herzenberg 1993).

In June 1944, as the Met Lab shifted to a support role for Hanford and Oak Ridge, work at the lab became less challenging. Both Karles submitted their resignations (Kathren et al. 1994). After six months on the project, Isabella returned to the University of Michigan as an instructor in chemistry.

One woman working at the Met Lab wrote, "I can only speak for myself of course, but while I was working there, I never encountered any discrimination" (Fineman 1996). Not all women working on the Manhattan Project were so lucky (Horning 1993; Howes and Herzenberg 1993; Herzenberg and Howes 1993; Rossiter 1995; Wise and Wise 1994). A case in point is that of Hoylande Denune Young, who had considerably more experience than Karle when she joined the Met Lab. Young, described by a colleague as "very forthright, quick-spoken, with a blazing intellect," was plagued by discrimination throughout her career as a chemist.

In high school, Young was allowed into the rigorous chemistry course for boys only because she could not fit the girls' course into her schedule. Once in the course, she was made to sit in the back row. The distance didn't prevent her from grasping the subject and deciding to make it her career. Young received a B.S. in chemistry in 1922 from Ohio State University and a Ph.D. in organic chemistry from the University of Chicago in 1926. As a graduate student of the chemist Julius Steiglitz, she completed a complex dissertation on stereoisomeric bromoimino ketones, which involved hand separation of individual crystals and the distillation of bromine (*American Men and Women of Science*; O'Neill 1979; Nicholls 1978).

Young began her working career as a research chemist in the lacquer industry at the Van Schaack Brothers Chemical Works on the North Side of Chicago. After four years, she accepted a position as an assistant professor of chemistry, teaching nutrition and biochemistry, at the College of Industrial Arts of Texas State College for Women in Denton, Texas. Texas proved a less congenial environment than Chicago, and Young came home when she was offered a research job at Michael Reese Hospital. Upon her arrival, however, the hospital director learned that she was a woman and refused to hire her (*American Men and Women of Science*; Kathren et al. 1994; Nicholls 1978).

Jobless during the Great Depression, Young accepted a position as an industrial chemist with the Pure Oil Company in 1938, where she worked until 1942 (Kathren et al. 1994). Again she ran into difficulties:

She agreed to collaborate with Cary Wagner of the Pure Oil Company in writing a book on petroleum refining. Having placed a rigid eighteen-month limit on her literature search, she found she was faced with another hurdle. Illinois law prohibited women from working more than forty hours a week. It was with great difficulty that she convinced Pure Oil that she needed a key to her office! The project lasted for six years but was interrupted by World War II and her book was never published. (Nicholls 1978)

In 1942, the University of Chicago offered Young a civilian position as a scientific librarian with the Office of Scientific Research and Development at the University of Chicago Toxicity Laboratory, a site for chemical-warfare research (*American Men and Women of Science*; Noyes 1948; Nicholls 1978). As chemical librarian, Young played an important role in assembling information from numerous British and American reports and in setting up the "master index" of toxic compounds, copies of which were placed in American, British, and Canadian chemical-warfare laboratories (Noyes 1948). She was one of several female scientists assigned by the OSRD to work as librarians, technical aides, and assistants—their talents grossly underutilized (Rossiter 1995).

Reassigned to the Met Lab, Young only later discovered that she was contributing to a top-secret atomic project (Nicholls 1978). As a senior chemist on the plutonium project in 1945–46, she edited papers that were later published in the *National Nuclear Energy Series*, the Atomic Energy Commission's report of wartime research on nuclear energy, of which Young was general editor (O'Neill 1979). She participated in meetings of the laboratory council and many of the technical meetings (*American Men and Women of Science*; Kathren et al. 1994).

Like Young, Ethaline Cortelyou started her career as a chemist and then became an editor on the Manhattan Project. After she received a B.S. in chemistry from Alfred College in 1932, she taught chemistry at several colleges and worked as a research chemist at Kingsbury Ordnance Plant and Indiana Steel Products before joining the Met Lab. In 1945, she was a junior chemist in the Information Division, where she worked as an editor on the University of Chicago campus. Among other activities, she worked with Helen Seaborg and Kathleen Hughes on preparing the classified table of isotopes, which became an essential item in project labs.

Cortelyou later became known as a vocal advocate for women's jobs in technical writing, editing, and journalism. She worked as an associate chemist at Argonne National Laboratory from 1951 to 1952, then as Chief Technical Editor at Aeroproject Inc. She was a Science Information Specialist at NASA in Washington, was Director of Volunteer Services for the National Institute of Arthritis and Metabolic Diseases, and taught at Wilson College in Pennsylvania (*American Men and Women of Science*; Kathren et al. 1994; Rossiter 1995).

In forming his team to study plutonium at the Met Lab, Glenn Seaborg relied heavily on friends and former students at Berkeley. Nathalie Seifert Baumbach graduated from UCLA, where she and her husband, Harlan Baumbach, were chemistry majors and classmates of Seaborg. After graduation they remained in California, where Harlan worked for Paramount Pictures, until Seaborg persuaded them to come to Chicago in the spring of 1943.

Nathalie Baumbach worked as a research assistant on the early microchemical studies of plutonium (Kathren et al. 1994). She was a member of the group that just a few months earlier had succeeded in isolating plutonium. In January 1944, she transferred to the group that was studying the recovery of plutonium from solutions. Seaborg specifically refers to her as a chemist. Her title, however, was research assistant rather than associate chemist or junior chemist. Her husband, who had received an M.A. in chemistry, had an appointment with the title of chemist. In August 1944, as the research at the Met Lab wound down, the Baumbachs returned to California (Kathren et al. 1994).

A number of other women worked at the Met Lab as research assistants and laboratory technicians in chemistry. What little we know of them comes mainly from Glenn Seaborg's extensive journals. Amanda Pauline Blume Stein received a B.A. in chemistry from Minnesota's Carleton College in 1944. She became a research assistant in the group studying the purification and analysis of plutonium compounds, and worked with solvent-extraction methods of purification. She also studied the distribution of light elements in systems extracting plutonium. She remained with the Met Lab until February 1945 (Kathren et al. 1994).

Roberta Wagner worked as a research assistant in the Solvent-Extraction Group in April 1946. Madeline Sohn worked very briefly in the Recovery Group, resigning after two weeks because of her fear of health hazards. She was replaced by Sylvia (sometimes spelled Silvia) Warshaw, who came to the Met Lab from the Tennessee Eastman Corporation in Oak Ridge, Tennessee. Warshaw, who had been born in Poland in 1921, studied at Wright Junior College in 1940–42 and the University of Pittsburgh in 1943–44. She worked at the Gulf Research Corporation and the Tennessee Eastman Company before coming to the Met Lab in January 1946.

Marilyn Howe worked as a research assistant in the Technical Division, where her husband, John Howe, was a group leader. The Technical Division was the section of the Met Lab primarily concerned with engineering and metallurgy. Virginia Meschke, a chemist at the Met Lab's New Chemistry Building, who showed the biologists how to separate out plutonium for injection into animals, was hired as a technician in the Solvent-Extraction Group in May 1946 (Kathren et al. 1994; Kisieleski).

## FEMALE CHEMISTS AND CHEMISTRY AT CLINTON ENGINEER WORKS

As the Manhattan Project expanded, it would have to incorporate new industrial processes and deal with radioactivity on a previously unheard-of scale. The Manhattan Project began to search for a site for both the planned industrial-scale uranium isotope-separation plants and pilot plutonium production and separation facilities. The project needed an isolated, inland site in a thinly populated area with abundant water and electric power. A site near Oak Ridge, Tennessee—within a reasonable distance of both the University of Chicago and Columbia University, where supporting research would be done—was selected in September 1942 as the location for the Clinton Engineer Works.

Because of the uncertainty as to which separation process (if any) would succeed, the Manhattan Project simultaneously pursued three methods to produce the fissionable isotope of uranium: electromagnetic separation, gaseous diffusion, and thermal diffusion. Plants for all three processes were built at Oak Ridge. Y-12 was the code name for the plant that produced U-235 by electromagnetic separation of uranium isotopes. The gaseous-diffusion plant was K-25, and the thermal-diffusion plant was S-50 (Graf 1994). X-10 referred to the experimental plutonium production reactor.

The Met Lab scientists considered themselves fully capable of constructing and operating industrial facilities, but General Groves wisely decided to give the job to experienced industrial contractors. In January 1943, he made initial arrangements with Du Pont to start construction of the pilot plutonium production facilities at the X-10 site. Faced with a revolt among the scientists, without whom the project could not move forward, the general agreed that the Met Lab would supply managers while Du Pont did the construction (U.S. Department of Energy/Martin Marietta 1992).

Union Carbide operated the gaseous-diffusion plant from its inception. K-25 was then the world's longest combined physics and chemistry process under one roof. It was a mile long and 1,000 feet wide, containing thousands of filters perforated with microscopic holes. The problem of fabricating barriers that resisted corrosion proved one of the major challenges to chemists at Columbia, the Met Lab, and Clinton, and in the companies involved in the construction of K-25. Uranium hexafluoride, the only uranium compound that was a gas at room temperature, was highly corrosive and reacted violently with any kind of grease or other organic substance. Nickel turned out to be the only metal that resisted the corrosion satisfactorily, so nickel and nickel-plated components were used throughout the facility. Even the pipes had to be nickel-plated (Groueff 1967).

Plutonium was produced by neutron absorption in the uranium fuel elements of atomic piles. These fuel elements, however, also contained uranium and radioactive fission fragments. Finding a safe way to separate plutonium from the fuel elements challenged the chemists and chemical engineers to design, fabricate, and test equipment for remotely transferring and evaporating liquids, dissolving and separating solids, and handling toxic gases. They needed instruments for remote measurements of volumes, densities, and temperatures in a hazardous environment. The chemists invented techniques to separate microscopic amounts of radioactive elements from volumes of liquid thousands of times greater. The unknown effects of intense radiation on the solvents had to be identified and dealt with. Project chemists had to dispose of contaminated equipment and unprecedented amounts of radioactive waste (*Oak Ridge National Laboratory Review* 1992).

A number of chemists from the Met Lab in Chicago moved to Oak Ridge in October 1943. There they continued their investigations of plutonium separation and the properties of plutonium compounds. The Clinton experience suggested that the bismuth-phosphate carrier process was not entirely suitable for plutonium separation, but a process using lanthanum fluoride worked well.

The Oak Ridge chemists produced minor miracles. In February 1944, the first plutonium shipments went from Oak Ridge to Los Alamos. By spring, Oak Ridge chemists had improved the bismuth-phosphate separation process to the point that 90 percent of the plutonium in the irradiated slugs was recovered. In early 1945, when plutonium separation ceased at X-10, the graphite reactor and separations plant had produced a total of 326 grams of plutonium, a substantial contribution to nuclear research and weapons development (*Oak Ridge National Laboratory Review* 1992).

Many women participated in the scientific work supporting both the uranium isotope-separation projects and the plutonium work in Tennessee.

Marion Crenshall Monet, a chemical engineer, received a B.A. from the University of Kansas in 1941. She studied chemical engineering at the Massachusetts Institute of Technology, where she earned her M.S. in 1943. Du Pont employed her as a junior engineer in 1943 and 1944. She and her husband, also a chemical engineer, then worked for the Manhattan Project at Clinton from 1944 to 1946 (*American Men and Women of Science*).

With a B.A. from Vassar and an M.S. from the University of Chicago in 1943, Marian Elliott Koshland was quickly drawn into the war effort. She worked as an assistant in the cholera project of the Office of Scientific Research and Development in Chicago in 1943 and 1944–1945, and then as a junior chemist on the atomic-bomb project in the Manhattan District in Tennessee in 1945 and 1946. In 1945, she married Daniel Koshland, a biochemist who was also working at Oak Ridge. She earned a Ph.D. in bacteriology from the University of Chicago in 1949, and became an immunologist on the faculty of the University of California at Berkeley (*American Men and Women of Science*).

Ellen Cleminshaw Weaver was born in Oberlin, Ohio, in 1925. In 1944 she married Harry Edward Weaver Jr., an experimental physicist, and followed him to Oak Ridge in June 1945. Lacking two courses for a bachelor's degree in chemistry from Flora Stone Mather College of Western Reserve University, she took summer courses at the University of Tennessee to complete her degree and graduated from Western Reserve University with an A.B. in 1945. At Clinton she worked on such projects as the study of fission fragments from the atomic pile and the development of microanalytical techniques for separating rare earths

(*American Men and Women of Science*; Howes and Herzenberg 1993; Weaver 1991).

Weaver—like the female janitors but unlike any male college graduate at the facility—had to punch a time clock. Men with her qualifications were paid a salary rather than an hourly wage, and their take-home pay was about twice hers. "I was paid 70 cents an hour, which was low even in those days," Weaver says. "It was insulting. And I got docked an hour's pay if I was as much as one minute late." Other women were treated similarly.

Weaver decided to protest to her supervisor, whose response was not encouraging: "My superior told me, look, this is the way it is, and if you don't like it, you can quit. Everyone thought women should be grateful just to be allowed to work. Equal pay was a completely foreign concept" (Horning 1993). But Weaver was determined, and she worked her way up the chain of command until she reached the president of Monsanto Chemical Company, which operated the part of Oak Ridge in which she was employed. The president of Monsanto immediately agreed with her and promised her a salary of $200 a month, but the Army vetoed it because the increase would have been too large a percentage of her pay. The attitude of the time was that *of course* women were paid less than men (Herzenberg and Howes 1993; Horning 1993).

As in the case of the Met Lab, we know that a number of other female chemists worked at Clinton, but we know little about them. Juanita "Billie" Wagner, who was the first woman to obtain a Ph.D. in chemistry from Brown University, worked as an analytical chemist at Oak Ridge. Other chemists included Helen Landriani Bench; Virginia Spivey Coleman, a chemist at Y-12 (Adamson 1994); Evelyn Kalichman Hanig; Emily Leyshon, also at Y-12; Grace Priest; Helen Stark; and Roberta Shor (Kathren et al. 1994). Kay Harvey, a lab assistant, transferred to Clinton from the Met Lab.

## FEMALE CHEMISTS ASSOCIATED WITH HANFORD ENGINEER WORKS

Producing sufficient plutonium for weapons required large reactors operating at high power and industrial-scale chemical plutonium-separation facilities. Many of the necessary devices had not yet been invented, but the risks of the processes clearly eliminated a site near

Chicago. Even Clinton lay too near population centers for comfort. The Manhattan Project began another search for a remote site with ample water and electric power. Construction started in January 1943 at Hanford, a large, arid tract in eastern Washington State, bounded on one side by the Columbia River. The facilities on the 500,000-acre site included three water-cooled production reactors to make plutonium and two (later four) enormous separation plants. Plutonium production at Hanford began in September 1944.

The Met Lab analyzed various possible designs for the production reactors. Simply scaling up Fermi's pile was not practical, because extracting plutonium from the uranium would have required disassembling the pile to remove the irradiated uranium—an inefficient, time-consuming, and dangerous procedure. Other potential designs included gas-cooled reactors using helium or air, water-cooled reactors, and reactors cooled by liquid metal. Du Pont selected a water-cooled design fueled by uranium rods enclosed in aluminum to protect them from corrosion.

Building the separation plants posed an exceptional challenge, even by Manhattan Project standards. It was daunting enough to scale up the size of the piles ten to twenty times while following a new design; the separation plants had to be scaled up roughly a billionfold from the lab experiments (Gosling 1990). Fortunately, the reactor and plutonium-separation work at Clinton furnished some intermediate steps. The conventional relationship between pilot plant and production plant existed between the Clinton labs' hot cells and similar concrete structures built at Hanford. The lanthanum fluoride process for plutonium separation was incorporated into Hanford's separation facilities (*Oak Ridge National Laboratory Review* 1992). Hanford could also draw on the experience of hundreds of people trained at Clinton, and many others from the Met Lab.

At its peak in 1944, Hanford Engineer Works employed approximately 4,000 women. Most of them were secretaries, clerks, nurses, food-service workers, barracks employees, and other support workers. The professional women were usually nurses. Most of the women were very young, often on their first jobs away from home. There was a small contingent of Women's Army Corps members. Less than 1 percent of the Hanford managers and supervisors were women (Gerber 1993).

Much of the research that specifically addressed the needs of the Hanford site was actually conducted at other locations, particularly at the Met Lab. There, a very unusual female metallurgist helped to solve the vexing problem of the corrosion of the aluminum coatings that separated uranium fuel from cooling water in the Hanford reactors.

A former colleague describes Nathalie Michel Goldowski as a "hotshot" metallurgist, and says that her solution to the problem of bonding the uranium slugs to the aluminum shells was a most important contribution to the project—one that "made it all work" (Freedman 1991). Goldowski was an interesting, intellectually uninhibited, flamboyant woman (Ringo 1991). Large but very graceful, she wore her heavy black hair down her back. Her clothes, which came from Paris, were hardly in keeping with the wartime Chicago norm. The purchaser of the car she used during the war remarked that it must have been one of her experiments in corrosion. Goldowski and her mother lived with several dogs that responded only to commands in French (Bernstein 1992).

Goldowski was born in Moscow into the Russian aristocracy in 1908. In 1917, she and her mother fled to Paris to escape the Russian Revolution. She received a Dr.Sc. from the University of Paris in 1935. Her doctoral work centered on the corrosion of metals. Having obtained a Ph.D. in physical chemistry in 1939, she went to work for the French Air Ministry and became Chief of Metallurgical Development at the age of thirty-two. As part of her work there, Goldowski suggested bolting strips of magnesium to the fuselages of French seaplanes, protecting the aluminum airframes from corrosion by salt water (Creutz 1992; *American Men and Women of Science*; Herzenberg and Howes 1993).

When Hitler occupied France, Goldowski and her mother again escaped, this time to the United States. She found work as a research associate for Sciaky Brothers in Chicago, and joined the Met Lab in 1943 (*American Men and Women of Science*). The small group founded by Leo Szilard did pioneering work on the metallurgy of uranium. (Gossip had it that the Metallurgy Group also had the only accident involving uranium at the Met Lab: A billet of uranium fell on a group member's toe.) In the ramshackle facilities of "the brewery," the Metallurgy Group operated both research labs and training facilities for Du Pont employees heading out to build Hanford (G. Sacher 1952).

Goldowski's areas of specialization included corrosion of metals, surface films on aluminum, stress corrosion, protection of aluminum alloys, and surface preparation of aluminum alloys for spot welding (*American Men and Women of Science*). Her work in the Metallurgical Laboratory dealt largely with corrosion of aluminum alloys and beryllium (Goldowski 1945). Uranium corroded badly when exposed to cooling water, and so, unfortunately, did the aluminum covers for the slugs (Brown and MacDonald 1977). The noncorroding aluminum coating developed by Goldowski was critical to the success of the plutonium production project.

## FEMALE CHEMISTS AND CHEMISTRY AT LOS ALAMOS

In early 1942, project leaders decided to consolidate the final research, design, and assembly of a bomb at one site. One week before Fermi's pile in Chicago produced its first sustained chain reaction, the Under Secretary of War directed that a site at Los Alamos, New Mexico, be acquired for a nuclear-weapons laboratory (Los Alamos Scientific Laboratory).

Chemists at Los Alamos were responsible for the refinement of plutonium and uranium-235 and for the preparation of metal for constructing weapons. They had to purify plutonium and uranium by removing lighter elements, as these might react with alpha particles from radioactive decay to produce neutrons, and such neutrons could cause a bomb to detonate prematurely. Fortunately, purification standards for uranium were relatively modest, and the Chemical Division at Los Alamos was able to focus on plutonium, about which less was known. Substantial progress was made on a multistep precipitation process for plutonium by the summer of 1944 (Gosling 1990).

The first small quantities of plutonium, in the form of a nitrate rather than metal, arrived at Los Alamos in October 1943. Early 1944 saw deliveries of several grams at a time, and larger amounts became available soon after, first from Oak Ridge and later from Hanford. In September 1944, the first kilogram of highly enriched uranium (63 percent U-235) was received from the isotope-separation facilities at Oak Ridge. By July 1945, fifty kilograms of enriched uranium had been received at Los Alamos, and the enrichment had increased to 89 percent.

Los Alamos not only recruited women to work at the site, but also made use of the wives of male employees. Lilli Hornig, a chemist, was born in Czechoslovakia in 1921, but subsequently became a U.S. citizen. She received an A.B. from Bryn Mawr College in 1942 and an M.A. from Harvard University in 1943 (*American Men and Women of Science*). She married Donald Hornig in 1943, the year he completed his Ph.D. in physical chemistry at Harvard. Donald Hornig was working at the Underwater Explosives Research Laboratory in Woods Hole, Massachusetts, when he was invited by the laboratory director to take a job doing unspecified work at an unspecified location. When he declined, pressure was brought to bear by other scientists, including James Conant, president of Harvard and a leader of war-related research, and George Kistiakowsky, head of the Explosives Division at Los Alamos, who called from Albuquerque. Hornig soon agreed to take the new job (Lamont 1965).

Lilli Hornig had almost completed her Ph.D. in chemistry at Harvard, but she decided to accompany her husband to Los Alamos. Within two weeks they had sold the yawl they had enjoyed sailing and bought an ancient Ford with frayed tires (Lamont 1965). They arranged for their household goods to be shipped after them, loaded their "new" car with personal belongings, and chugged off to New Mexico.

Their arrival at Los Alamos was inauspicious. Lilli Hornig took one look at the primitive living conditions and the cheerless barracks that would be their first home on "the Hill" and burst into tears (Lamont 1965). She was then asked to take a typing test. She had a master's degree in chemistry from Harvard, but typing had not been one of the requirements. Fortunately, someone recognized the error, and she was assigned as a staff scientist to a group dealing with plutonium chemistry, where she worked with one of the few other women with the white badge marking a scientist, a chemist named Mary Nachtrieb.

Plutonium purification efforts were curtailed in 1944 when plutonium-240 was discovered, and Hornig and Nachtrieb's work ended abruptly. Hornig transferred to the Explosives Division, where she led a section dealing with the development of explosive lenses (Henriksen 1986; Hornig 1993; Hoddeson et al. 1993; *American Men and Women of Science*). Her husband, who later served as a science adviser to President Lyndon Johnson, described her job transition: "Then at some point they decided plutonium chemistry was too dangerous for women, so she went on to high-explosive lenses instead—which was a little crazy" (Borman 1995, p. 54).

The work on high explosives supported the development of a plutonium weapon. In the uranium bomb, a critical mass was achieved by firing one piece of uranium metal into another. For a plutonium bomb, a sphere of plutonium was uniformly compressed using conventional explosives, bringing the plutonium to a high enough density to create a nuclear explosion. The implosion bomb would work effectively only if the entire surface of the sphere were compressed to precisely the same degree at precisely the same time. Early attempts at creating implosions showed that they were far from symmetrical, with some points on the bomb's surface being compressed sooner than others. The time difference might only be millionths of a second, but that was enough to cause failure.

The creation of a sufficiently uniform implosion was accomplished by using explosive "lenses," which changed the shape of detonation waves the way glass optical lenses change the shape of light waves. The lenses consisted of shaped pieces of different types of explosives in which detonation waves traveled at different speeds. One of the biggest uncertainties about the plutonium bomb was how well the explosive lenses would actually work (Borman 1995).

After August 1944, the program for exploring and developing a plutonium implosion weapon absorbed an increasing fraction of Los Alamos's personnel. Hornig's work included fabricating lenses in devices that reminded her of cookie molds. The explosives were detonated, and their performance was determined from an imprint on a steel plate. Hornig also used X-ray photography to study the detailed progress of an implosion. No available theory was precise enough to guide the group's efforts, so design proceeded by trial and error (Hoddeson et al. 1993).

Chemists, among them Norma Gross, who arrived at Los Alamos in 1944 as a WAC, played a critical role in a second method for assessing the progress of an implosion. Radioactive lanthanum-140, which emitted gamma rays, was seeded in the metal core. As the explosion compressed the center of the system, the density of the system increased and fewer gamma rays could penetrate the imploding mass to reach the ionization chambers placed around the test explosion. By monitoring the current in the chambers, scientists could create a continuous record of the collapse of the core.

Radioactive lanthanum-140 had a half-life of only forty hours, but it was formed by the decay of barium-140, a common fission fragment with a half-life of twelve and a half days. Barium was separated from other fission fragments in a special "hot" laboratory at Oak Ridge and shipped to Los Alamos, where the chemists separated the lanthanum so that only the shorter-lived isotope would contaminate the environment of the explosion. The explosions took place in Bayo Canyon, well away from the main lab, in a mobile setup using old military combat tanks fitted with the necessary instruments. The problem facing Gross and her colleagues was handling the extremely radioactive barium and concentrating nearly 100 curies of radioactive lanthanum in a tiny cone. The cone was then transported to the canyon and lowered into the assembly using a ten-foot-long rod like a fishing pole (Hoddeson et al. 1993).

Gross held a master's in chemistry from Bryn Mawr and had worked on several projects for the New York University medical school, designing procedures to adapt older medical tests to new electronic instruments. Though she says that chemists were an "inferior species" at Los Alamos, she received the white badge that allowed her to attend all the colloquiums at the lab and thus to stay informed of the progress of the atomic bomb.

When shipments of radioactive barium and lanthanum arrived at Los Alamos, the chemists, under the direction of Gerhardt Friedlander, immediately began separation procedures using a long and complicated vacuum system. The procedure required careful monitoring into the night. Gross found herself at an advantage because she could stay awake more easily than most members of her group. She also found that handling the radioactive cones was easier for her than for her colleagues. At five feet tall, she could walk upright under the protective shielding to manipulate the extremely radioactive samples, whereas her taller male colleagues had to contort themselves to escape a large dose of radiation. The group took turns handling the material to minimize any single person's risk (Gross 1997). By June 1945, Gross and her colleague Rod Spence had developed a new procedure for separating the lanthanum from the barium: precipitation of lanthanum hydroxide. A great advantage of this process was that it could be carried out remotely, from ninety feet away (Hoddeson et al. 1993).

Like many other women, Norma Gross was an unofficial observer of the first nuclear test. With her husband, an electronics specialist and Army officer who had been recruited by Kistiakowsky to work on electronic monitoring of weapons effects, Gross drove to a wooded site with a view of ground zero and camped out through the long night. She claims that all the maintenance workers at the lab knew where and when the test would occur, so there was no problem getting there (Gross 1997).

In other areas, security proved more effective. Gross's father traveled to Colorado to see her brother and called asking whether he could come to visit her. Despite his protests, she refused to see him, saying that she was "too busy." He was surprised to find that the operator would not charge him for the call, and decided that whatever his daughter was doing for the war must be important.

Elizabeth Rona, a radiochemist, never received an official appointment on the Manhattan Project because she had been born in Budapest. Nevertheless, being a world-class expert on the radioactive element polonium, she made a significant contribution. Rona's father was one of the first physicians in Hungary to take an interest in the medical applications of radium. She received a B.A. at the University of Budapest, and earned her Ph.D. in chemistry, physics, and geophysics at the age of twenty-one. She began research on the disintegration series for radium around 1919, while working as an instructor in biochemistry in Budapest (Tennenbaum 1995; Rona 1978).

In 1921 Rona moved to the Technical University in Karlsruhe, Germany, where she worked under George Bredig, a leading physical chemist. From time to time Bredig invited students to his house for tea. Rona reminisced: "The only drawback was that I had to join the ladies, being a woman, the only one at the time in the laboratory. I was much out of place. The conversation was about children, cooking, preserves. Recipes were exchanged; to these I could not contribute. How I longed to be with my colleagues, to hear and talk shop" (CWP 1996).

In 1922, Rona became a fellow of the Kaiser Wilhelm Institute in Berlin and worked there for two years with Otto Hahn and Lise Meitner (Tennenbaum 1995). In 1924 she received an appointment at the Vienna Radium Institute, remaining there until 1938 (*American Men and Women of Science*). In 1928 she visited Paris and developed her expertise on polonium by working with Irène Joliot-Curie. She also spent time at the Cavendish laboratory in Cambridge, England (Rona 1978).

Rona emigrated to the United States in 1941, and later became a naturalized citizen (Tennenbaum 1995), but her early experiences were hardly welcoming:

> One of my first trips to New York was to Columbia University, where I knew that some of my former colleagues did some kind of research. I went straight to the Physics Department. I saw Fermi, Szilard, and quite a few whom I had met previously. I had the quaint impression that they

did not know me. Nobody talked to me. A scientist, formerly from Vienna, to whom I brought special greetings from his former professor, turned his back and fast departed. Many years later Professor John Dunning, the only scientist who talked and was kind to me, explained the situation. Men from the FBI were present all the time, questioning them about what I, a person with a visitor's visa, was doing at the laboratory where such secret work is done. Surely I would go back to Germany and report some secret results. (CWP)

Rona worked at the Geophysical Laboratory in Washington, D.C., in 1941, and was an associate professor at Trinity College from 1941 to 1946 (*American Men and Women of Science*). While there, she received an unexpected invitation to join the war effort:

While at Trinity, I received an urgent call from an important chemical company, requesting that I come for an interview at my earliest convenience, which I did. They did not tell me what they wanted me to do; from the questions I understood that I was supposed to do some radioactive work, especially with polonium. I found out later that they did intensive work for the war effort, especially on polonium. Nothing came from this interview, however; my visitor's visa, with my family still in Hungary, was too much of an obstacle for such employment. (Rona 1978, pp. 56–57)

But somewhat later, when Rona returned to Trinity College after spending the summer in Los Altos, California, she was handed a telegram from Brian O'Bryen, a professor at the University of Rochester's institute of optics. It was marked "restricted" and read as follows:

In connection with a certain war work, immediate need has arisen for large quantities of polonium, and probably also for lead-210. A stockpile of radon seeds will be available. At first, it will be necessary to produce these elements in amounts corresponding to about 50 milligrams of radium, and it is desired to obtain this quantity in the shortest period of time. It is probable that considerably larger amounts will be needed thereafter. Solutions of polonium-210 and lead-210 should be without contamination with inactive material. It is also desirable that these solutions should be strong and either not at all or slightly acid. I believe that your unusual experience in radiochemistry will ensure quick and reliable results. You would be making a substantial contribution to the war effort. (Rona 1978, p. 56)

Polonium, on which Elizabeth Rona was the world's leading expert, was discovered by Marie Curie. It is a peculiar metal—as soft as cream cheese, dangerous to inhale, and difficult to produce. Until the time of the Manhattan Project, it had been produced only in laboratory quantities. The element was not classified at the time of O'Bryen's telegram, though it was later classified and given the code name "postum" (*Los Alamos Science* 1995).

Rona describes her response to the telegram:

> I replied immediately that I would be interested in that assignment. At 11 p.m. the next day there was a knock on my door. A tall, thin gentleman introduced himself as Professor Brian O'Bryen. He explained the details of my work and suggested that one of the students should help me in some of the manual work, but that she should be neither a physics nor a chemistry student, in order that she would be quite oblivious to the nature of her work. (Rona 1978)

Polonium became important to the Manhattan Project in several connections. The element is radioactive and emits alpha particles as it decays. If alpha particles collide with beryllium atoms, neutrons are ejected. In a plutonium bomb, at the moment of implosion, neutrons are needed to ignite the fission chain reaction. Spontaneous fissions do occur, but the chance of enough of them occurring at precisely the right time is too small to depend on. The device that provides neutrons at precisely the correct time is called an initiator. The initiators, nut-size devices of beryllium and polonium, were placed between the two hemispheres of the bomb. The implosion would mix the polonium and the beryllium, producing neutrons that would ignite the chain reaction (Groueff 1967; Hoddeson et al. 1993).

Rona referred to her polonium work as highly confidential, kept its nature secret from her assistant, and even in her memoirs, published in 1978, treated the information with what now appears to be excessive discretion. She explains:

> O'Bryen must have been satisfied with my work, because at the end of the school year he called on me to come to Rochester to try to solve a difficulty that arose in work for the Office of Scientific Research and Development (OSRD). I was not yet a U.S. citizen, and the work was highly confidential. Soon I found out that O'Bryen did not allow difficulties to interfere with the completion of my assignment. After a short vacation I got word that my security clearance had come through. I soon solved the difficulty and explained the procedure to a representative of the Canadian Radium and Uranium Company, which had contracted to do a mass production job for the OSRD. I was asked if I would give away my method without selling it.

I gave all the details without compensation. The method then went into mass production. Part of the gadget, the metascope, was a zinc sulfide screen, with the best characteristics of a scintillation screen. I was asked to spend the rest of the summer working on developing the metascope. It took several months, but finally I succeeded. (Rona 1978, p. 57)

Polonium posed new health risks to Manhattan Project workers. It seems highly likely that the polonium that Rona prepared and the methods that she developed for preparing it were used in early studies of the biological effects of the element, and perhaps in other applications, including initiators. The University of Rochester undertook human and other biological studies using polonium as part of the Manhattan Project. Documents bearing Elizabeth Rona's name are included in archives of the government's Office of Human Radiation Experiments (OHRE 1996).

A description of the Rochester work in the journal *Los Alamos Science* closes by noting: "Such metabolic studies were possible at Rochester University in 1944 because polonium was available at that time. The research yielded important information for the Manhattan Project on the hazards of polonium and helped develop techniques for the similar but later studies of plutonium" (*Los Alamos Science* 1995).

In 1947, Rona went to work at Argonne National Laboratory. She tells a story from her days at Argonne: "An amusing episode occurred ... during my stay at Argonne National Laboratory. A biologist from the Medical Division needed a polonium source for his single-cell experiments. Knowing he could get a polonium source from Dr. Rona, he asked his colleagues where in Europe she could be contacted. He was told, 'She is in the next room'" (Rona 1978, p. 28).

In 1950 Rona began work at the Oak Ridge Institute of Nuclear Studies (now Oak Ridge Associated Universities), and she subsequently worked at the Institute of Marine Science of the University of Miami (Tennenbaum 1995; *American Men and Women of Science*; Rona 1978). She died in 1982 at the age of ninety-two (Tennenbaum 1995).

Other female chemists contributed their expertise to the Manhattan Project in various ways. Margaret Foster received a B.A. in chemistry from Illinois College. She joined the U.S. Geological Survey in 1918 as an analyst of water resources, and became the organization's first female chemist. She worked for many years in the Water Resources Laboratory, where she perfected methods for detecting minerals in natural waters. She wrote several papers on groundwater in the coastal plain formations of the Gulf Coast States. During World War II, Foster devised two new methods for uranium and thorium analysis for the Manhattan Project, and two new ways of separating the two elements. When she returned to the Geological Survey, she switched her studies from groundwater to the geochemistry of clay minerals and micas (*American Men and Women of Science*; Kass-Simon 1990; Rossiter 1982).

The chemistry studied as part of the Manhattan Project covered numerous aspects of the discipline: the organic effects of radiation; instrument development; new microanalytical techniques; the development of the field of radiation chemistry; new studies of high explosives; the discovery of new isotopes; and more. The science needed to produce materials with necessary properties, both for manufacturing plants and for the interior of the bomb itself, set a global standard. Women worked in all aspects of this massive undertaking.

Most of the women enjoyed their work and speak proudly of the job they did. They recognize that the success of the Manhattan Project depended on the success of many smaller efforts. They see their success not as the triumph of a small group of brilliant scientists but as the culmination of the efforts of many hardworking, dedicated people. One of them put it this way, "I think one of the remarkable things also from a broad general point of view... was the idea that this enormous project was going to go forward with determination and earnestness. I said to myself, 'Good heavens, are we really going to do all that?' And we did it" (Friedell 1994, p. 34).

## 5  Mathematicians and Calculators

IN THE RUSH to beat the Nazis to the atomic bomb, the Manhattan Project scientists explored many possible solutions to various technical problems essentially at the same time. General Groves rushed processes from lab benches to full-scale production plants, leaping over the pilot-plant phase standard in most manufacturing projects. Tests of the bomb itself had to be largely omitted as well: Uranium-235 and plutonium were far too expensive. A uranium bomb—the type that was used at Hiroshima—was never tested at all.

Without pilot plants or weapons tests, the Manhattan Project had to rely heavily on numerical simulations. Computers had not yet been invented. Electric punched-card machines were available for certain calculations, but most problems were solved on heavy, old-fashioned desktop calculators. A mathematician would devise an iterative procedure for making a simulation. The operators of the machines would take the equations and punch numbers into their calculators over and over, performing a single operation or set of operations on a long series of numbers in a human analog to today's parallel processors.

These calculations provided a necessary foundation for the bomb, and female mathematicians and "calculators" provided essential expertise and patient labor in conducting them. Some histories have mentioned their work and listed some of their names. Few, however, have recognized the skill with which women organized the work necessary for this huge project in applied mathematics.

A number of talented women worked on the numerous problems of applied mathematics that accompanied almost every phase of the construction of the atomic bomb. The highest-ranking woman, Mina Rees, who held a Ph.D. in mathematics from the University of Chicago, served as executive assistant to the chief of the Applied Mathematics Panel in the OSRD. Her Washington job allowed her full access to data on the Manhattan Project and other wartime research efforts. At the other end of the spectrum, numerous women with liberal-arts degrees punched numbers into the desktop calculating machines without much idea of what they were doing.

Theoretical calculations, unlike experiments, were conducted around the country, so female mathematicians were widely scattered among different groups. In a typical case, Hilda Lefkowitz planned to complete her bachelor's degree in mathematics at Hunter College in New York in February 1943. (At Hunter, her upper-class "grandmother" was Rosalind Sussman—later Yalow—a physics major and a future Nobel laureate.) By November 1942, Lefkowitz had received job offers from a variety of institutions. But when a secret project at Columbia University sent letters to the placement offices at Barnard College, Brooklyn College, and Hunter, seeking interviews with their top two graduates in mathematics, Lefkowitz accepted the interview.

In the physics building at Columbia, she was impressed with the conviction of the men interviewing her that this work was extremely important. The small Columbia research effort on uranium had just received large-scale federal support and was growing rapidly. The project offered Lefkowitz a job for very little money; only Columbia, her physics professor at Hunter snorted, would have the arrogance to offer such a minuscule salary. Nevertheless, lured by the excitement of the classified work, Lefkowitz accepted. As part of the security check, she claims, the FBI interviewed her mother's butcher.

The final job offer arrived inside a lined envelope secured by a blue wax seal to prevent tampering. The project officials wanted Lefkowitz to start in December, and they persuaded Hunter to waive the rest of her semester classes, although she insisted on completing her final exams. She was assigned to Karl Cohen and Irving Kaplan, two physical chemists. Kaplan and Cohen would derive differential equations from their physical work and present them to Lefkowitz and Bernice Schwartz—the other mathematician recruited by Columbia, who came from Brooklyn College. The women, working with desktop calculators, would plug a series of numbers into their equations.

Lefkowitz considered it ironic that she should be working with differential equations, as her professors at Hunter had assured her that this was one subject she didn't need to take. She also remembers that the Columbia Physics Department, which had no women's rest rooms, hastily assigned rest rooms on alternate floors to women, a situation that remained unchanged until at least 1970. With the arrival of full federal funding in January 1943, the group expanded quickly and moved off the Columbia campus to the Nash Building on Broadway and 133rd Street, where security could be tightened. The fact that the building had been an automobile dealership lent it character.

As soon as the security checks on Lefkowitz had been completed, she was briefed on the nature of the project. Her understanding was that work on the bomb was very important because Germany might develop it and Russia certainly would. She also felt that it was unlikely that the weapon would be ready in her lifetime. Because she read a little German, Lefkowitz traveled to the Columbia campus each week to screen microfilmed lists of articles published in German technical journals, looking for anything that might be of use to the uranium project. Of her job, which involved a great deal of detailed computation, Lefkowitz says, "It didn't take much imagination, but you *had* to do it correctly."

She recalls the project staff's shock when a radio talk-show host announced that a top-secret project on upper Broadway was building a superbomb. The host was fired, and the project workers lied to their friends and relatives about the announcement. As construction started at Oak Ridge, many of the experimentalists left for Tennessee, but the theorists stayed in New York. Rudolph Peierls and Klaus Fuchs, "the British" in local parlance, paid frequent visits to the New York group.

For a while, the pace continued to be frantic and the work exciting, but by the spring of 1945 things had slowed down. Lefkowitz had been taking mathematics for engineering students at Columbia, where her professor recognized her talent and recommended her to Raymond Mindlin, who was on leave from Columbia to Bell Laboratories, then located in New York City. Mindlin had done outstanding work on developing the proximity fuse. With that work completed, he needed an assistant on his current projects. Lefkowitz was assigned to the Mathematics Department, but really functioned as Mindlin's personal assistant (H. Cooper 1996).

Not all mathematicians on the Manhattan Project had intended to study math. Joan R. Clark, who was born in Madison, Wisconsin, in 1920, worked as a junior science aide in physical chemistry in the Eastern Regional Research Laboratory of the U.S. Department of Agriculture. In 1943 she became a math assistant for the Manhattan Project while working for the Carbide and Carbon Chemical Corporation. After the Manhattan Project she became involved in crystallography. She received her Ph.D. in the subject from Johns Hopkins University in 1958 and eventually served as an investigator of the Apollo moon rocks (*American Men and Women of Science*).

In contrast, Ann Kenny majored in mathematics and wanted to attend the graduate program in math at Princeton University. Princeton was then, and remained well into the 1960s, a male-only institution. Kenny attended Columbia instead, and was making progress toward her doctorate when she was invited to Princeton to work on the Manhattan Project. She found the work enormously challenging, and said that she never needed an alarm clock because she was so eager to get to her office. Neighbors claimed to be able to set their watches by Kenny as she hurried to catch the bus to work.

Participation in the Manhattan Project cost Kenny her graduate degree in mathematics. On leaving New York for Princeton, she had packed her research materials into boxes in her father's apartment. He threw them out, thinking that they were only old papers. While trying to summon the strength to start over, Kenny was offered a job running the Math and Physics Library at Princeton. She accepted and continued working there until she retired. She maintained her own network with libraries around the world and was able to obtain papers from Eastern-bloc countries even at the height of the cold war. A colleague describes her as "a beautiful woman, a good and trusted friend and a brilliant mathematician" (O'Brien 1993).

Lefkowitz, Clark, and Kenny found themselves swept up by the Manhattan Project almost by accident. They did not join a mathematics section associated with the project but performed real mathematics in its service as part of some other group. There were other female mathematicians at diverse locations about whom we know little. Mathematicians are harder to track than bench scientists, who worked at a limited number of locations and often knew one another. Two mathematicians whose names we do know are Grace Estabrook, a statistician at Oak Ridge, and Ardis Monk, who worked at the Met Lab.

## THE "COMPUTERS" AT LOS ALAMOS

Before 1943, most of the work on the design and functioning of atomic bombs was theoretical, based on small-scale nuclear-physics and reactor experiments. But in that year a new laboratory, built specifically to design and create such a weapon, was established at Los Alamos. Unlike Hanford or Clinton, Los Alamos was actually an Army base. Once inside the closed city, scientists and their families were forbidden to leave without permission. Some women with degrees in math or physics were recruited to help with the extensive calculations needed for designing the bomb. Others, already on "the Hill" in their capacity as wives, joined the project as full- or part-time "calculators" or "computers," as the people who performed calculations on desktop machines were often called in those days.

The design of the implosion type of atomic bomb involved placing shaped conventional explosives around fissionable material in a configuration that would generate a uniform shock wave when detonated. The shock wave would compress the fissionable material to a critical state. The complete release of nuclear energy by fission would take place only after about fifty-seven fission generations. Each generation required about one "shake," the term given in Los Alamos to the time interval of ten nanoseconds—that is, ten billionths of a second. Most of the fission energy was released in the last few generations, so if the device blew itself apart before about fifty-seven shakes had elapsed, the detonation would be expected to release only a small fraction of the intended energy and the explosion would be a fizzle. A successful bomb would need extremely uniform compression, which would not be possible with explosives that did not burn uniformly or detonators that did not fire with nanosecond precision.

Robert Oppenheimer and his advisers initially ignored the implosion design in favor of the "gun" design, in which a critical mass was produced by shooting a plug of fissile material into a hole in a sphere, also made of fissile material. The gun had to shoot the plug at very high velocity, since fissions began to multiply as soon as the plug neared the sphere. If the chain reaction built up before the plug was fully inserted, the weapon would destroy itself without releasing most of its nuclear energy. Once the physicists realized that they could make the gun out of very light materials—it had to fire only one shot, after all—they made rapid progress on a gun that could detonate a fission explosion in uranium-235.

When the first samples of plutonium reached Los Alamos, the physicists soon recognized that no gun would be fast enough to construct a plutonium bomb. When fissile plutonium-239 is created in nuclear reactors, the reactor also produces a second isotope, plutonium-240, which fissions without absorbing a neutron and produces neutrons when it does fission. Adding neutrons from a sphere to those from an incoming plug of plutonium would cause an early start to the chain reaction, an event known euphemistically as predetonation. Predetonation would destroy the weapon before most of its nuclear energy could be released. Physicists had to find a faster way to assemble a critical mass of plutonium. They decided on implosion.

It was known from the outset at Los Alamos that the hydrodynamics of implosions and explosions could not be studied analytically but would have to be treated either experimentally or numerically. Complex calculations were required to trace the path of the incoming shock wave from the detonation of the explosives through the fissile core of the weapon, and the outgoing shock wave that produced the explosion. In the summer of 1944, project leaders committed themselves to constructing an implosion weapon using plutonium in case the project to enrich uranium failed—which seemed relatively likely as the isotope-separation projects hit snags. The models constructed by the Los Alamos computers were critical to assessing implosion designs and the feasibility of the implosion bomb itself.

Such calculations lend themselves well to modern computers, but not to desktop calculators that performed one mathematical step at a time. Some of these calculators were later replaced by IBM electric calculating machines, which received data automatically, on stacks of punched cards. The IBM machines speeded up the calculations greatly and proved crucial in calculating the behavior of imploding systems. The vast number of calculations needed for the creation of the atomic bomb also helped trigger the development of the first electronic computers, and the IBM machines were used to check the functioning of those computers (Metropolis and Nelson 1982).

The Theoretical (T) Division at Los Alamos struggled to develop the nuclear and hydrodynamic criteria for designing the atomic bomb and predicting its performance (Brown and McDonald 1977). Los Alamos was able to take advantage of the experience of Stanley Frankel and Eldred Nelson, two theoretical physicists who had organized a computing service for the electromagnetic isotope-separation project at the University of California at Berkeley the year before. They had also performed early calculations relating to the diffusion of neutrons in a critical assembly of uranium. When they came to Los Alamos in the spring of 1943, they ordered the same kind of machines that they had used in California: Marchant or Frieden desk calculators.

During initial operations at Los Alamos, these desk calculators—from Marchant, Frieden, or Monroe—were the only aids for doing the calculations. The calculators were hard to obtain, being in high demand around the country, and they required operators with some mathematical knowledge, the ability to pay attention to detail, and nearly infinite patience. Many of the staff members who performed the calculations were female, although soldiers from the Army's Special Engineering Detachment supplemented the staff of T Division. Among the early computing personnel at Los Alamos were Mary Frankel, Josephine Elliott, Beatrice Langer, Mici Teller, Jean Bacher, Kay Manley, Emma De Le Vin, Margaret Johnson, Frances Wilson (Frances Kurath), Bernice Brode, Edith Wright, and Betty Inglis (Metropolis and Nelson 1982; Libby 1979). Several of these women were the wives of prominent scientists. Some wives worked part-time because they had small children or because they wanted the better maid service available to women who worked in the labs. In the late summer of 1943, Donald Flanders, a mathematician from New York University, was hired to head this hand-computing group, known as T-5.

The computing demands of Los Alamos quickly outstripped T-5's capacity. In late 1943, one of the staff scientists suggested augmenting the T-5 desk-calculator operation with IBM punched-card machines. The T Division received permission to acquire the most up-to-date electromechanical IBM accounting equipment. Eventually the division established two groups, T-5 computations and T-6 IBM computations, which employed the new, faster machines.

Women were recruited for both computing groups. "We hire girls because they work better and they're cheaper," one division leader is rumored to have said. Not all women in early computing were concerned about being underpaid, however. "It all had to do with expectations," said one of the female programmers who worked on the Whirlwind computer at MIT around the end of World War II. "At that time, working women were expected to be nurses or schoolteachers. Thus, to be given the chance to work in a technical field was a great opportunity. However, upon closer inspection, almost all the leadership and managers were men" (Gurer 1995, p. 48). Similar observations were made by one of the early female programmers working with the ENIAC computer.

Joining the technical work was simple for wives, who underwent no additional security checks and were allowed to arrange their schedules to accommodate the demands of their families. Beatrice Langer was one who found out just how simple it was to join. Langer had obtained a degree in art from the New York School of Fine and Applied Art, but she sat in on mathematics courses at Indiana University when she moved there with her husband, a physicist. These math credentials were good enough for the scientists at Los Alamos: She was recruited to perform calculations as soon as she arrived (Langer 1996).

Mary Frankel, the wife of Stanley Frankel, supervised the women who operated the Marchant desk calculators in Donald Flanders's Theoretical Group. She had degrees in psychology and mathematics, and had worked as a psychologist before joining the Manhattan Project as a junior scientist at Los Alamos in 1943. She became an expert in the use of numerical methods to solve physical equations, and she set up the problems for the staff to run on the desk calculators. Frankel wrote out the problems in her small office and placed them in a basket labeled, "Free—take one." Young, blonde, and serious, she was a demanding boss to her crew of tired wives, most of whom were raising small children and working the standard three-eighths time. She urged them to study math at night in addition, and she felt free to let them know how ignorant they were.

Bernice Brode presents the following example of Frankel's approach: "'Mrs. B.,' she would say, 'there's a great future in this computing business if you could possibly learn decimals.' Mary was a good twenty years my junior, so I reminded her that *this* was my future. I once asked her why we worked three-eighths time and she replied, 'Mrs. B., if I could answer questions like that, well, I'd be in charge of this whole outfit'" (Brode 1980, p. 148). Considering her management style, it is hardly surprising that Frankel later had difficulty managing SEDs (Special Engineer Detachment GIs) and professional women assigned to the IBM machines.

The women whose experiences are described here must stand in for a number of others about whom we have not been able to learn as much. They are a sample of the kinds of women who labored at the endless calculations needed to field an atomic bomb.

Like many wives whose husbands joined the staff at Los Alamos when the laboratory was founded, Kathleen "Kay" Manley had graduated from college and done some graduate study. A Canadian, she obtained her undergraduate degree and teaching certificate from the University of British Columbia. She then pursued a master's degree in education at Teachers College of Columbia University. Teachers College lies just across 120th Street from the physics building, and the young teacher fell in love with a physicist named John Manley. They married and moved to the University of Illinois, where John had obtained a faculty position.

Shortly after Pearl Harbor, John Manley joined the Manhattan Project, quickly becoming Robert Oppenheimer's second-in-command. He moved to Los Alamos in April 1943 with the first contingent of scientists. Kay Manley remained in Chicago until June, then traveled west by train with her toddler and her eight-week-old baby. The reunited family spent a weekend at Bandelier National Monument in northern New Mexico, then headed for their new home at Los Alamos, where they found all their furniture piled in the middle of the living room. It took time and energy to straighten out the mess.

As soon as she had her family's life running smoothly, Kay Manley joined the technical workforce. General Groves considered it a waste of resources to maintain civilian personnel, and put pressure on all wives to work at least part-time. Maids were provided for child care, and the variety of talent the wives possessed was put to good use.

Manley was assigned to the Calculations Group (group T-5) directed by Donald "Moll" Flanders. Five or six women sat at tables equipped with Marchant calculating machines and processed ten- to fourteen-digit numbers according to instructions. Each machine stood about a foot high, occupied several square feet of desk space, and clanked and banged continuously. The noise led to frequent mistakes in the detailed work toward the end of a four-hour shift. The machines also malfunctioned fairly frequently, and tended to break down under the heavy use. The group ran each calculation twice, and a female supervisor checked them for errors. If her checks revealed a broken machine, it had to be shipped back to the manufacturer. Later, the group was able to call on its own repairman or, if the repairman was unavailable, the young theoretical physicist and future Nobel laureate Richard Feynman (Manley 1996).

Although many of the "calculators" came to Los Alamos with their husbands, the lab also recruited locally. Margaret L. Johnson grew up in Espanola, New Mexico, and obtained a B.S. in mathematics with a minor in home economics from the University of New Mexico in 1944. After a semester of graduate work, she felt it was time to go to work in support of the war effort. She dropped in at the Santa Fe office of the Manhattan Project—everyone in the area knew that there was secret war work going on, though they didn't know what kind—and asked whether the project needed someone with her background. Three weeks later, in August 1944, she joined the Los Alamos Theory Division. Her group, which was nearly all female, solved differential equations and other mathematics problems without understanding the physics behind them. They also made plots with french curves if they were needed. The available tools were Marchant calculators and a set of WPA mathematical tables that had been published years earlier (Johnson 1991).

Frances Wilson actually had formal training in mathematics. She received a B.S. degree in the field from Iowa State University, then worked for Blue Cross before learning of a job opportunity with the Manhattan Project. She came to Los Alamos in 1944, and worked there until 1946. She did hand calculations in group T-5, the same group as Betty Inglis, Mici Teller, Beatrice Langer, Emma de la Vin, Josephine Elliott, Margaret Johnson, and Edith Wright. (Some of the other women involved in calculations were in other groups; these included Naomi Livesay and Frances E. Noah, who were scientific assistants.) As group T-5 operated, scientists brought in problems they wanted solved, usually numerical integrations.

In May 1945, while at Los Alamos, Wilson married Dieter Kurath, a physicist. Margaret Johnson was "best woman" at the wedding. In May 1946, the Kuraths left Los Alamos and came to the University of Chicago. While Dieter pursued a graduate degree, Frances worked under the west stands doing opacity calculations in a group that also included the physicist Maria Mayer. Frances subsequently worked at Site A with Dave Inglis, a physicist and the husband of Betty Inglis; "Moll" Flanders; and Mort Hamermesh, a theoretical physicist who later worked at Argonne National Laboratory on classified calculations relating to reactors. Frances and Dieter later both worked in the Physics Division of Argonne, Frances as a computer scientist from October 1947 to September 1950, when she quit to have her first child (Kurath 1990).

Bernice Brode attended Occidental College (now UCLA), and received bachelor's and master's degrees from the University of California at Berkeley. She married Robert Brode, a Berkeley physics professor. Before the war, she traveled with her husband on his fellowships and visiting positions at Princeton, London, Cambridge, Manchester, MIT, and Washington, D.C. Early in World War II, she taught English at the Soviet military mission in Washington. She joined her husband at Los Alamos in September 1943 and stayed until December 1945.

On her first trip to Los Alamos, Brode was "bewitched by the scenery—the stretches of red earth and pink rocks with dark shrubbery scattered along the ochre cliffs; lavender vistas in the distant Sangre de Cristo mountain range. Color was everywhere. Occasionally adobe houses arose from the earth with strings of scarlet chili peppers hung outside to dry. The flat roofs were strung with ears of yellow, blue, white, and dark red corn, also drying in the sun for tortillas" (Manley 1990; Brode 1980).

Brode worked at Los Alamos as a "computer" during her entire stay there. She and other wives, hired for the work but not scientists or mathematicians, were trained by Joe Hirschfelder, a chemistry professor and ballistics expert. The computers worked in a large room adjacent to Hirschfelder's small office, where he tacked up *Esquire* calendars "for decoration and cheerfulness," as he explained. After three months of computer training, some of the wives moved to T Division to work in Flanders's Computation Group.

Though the calculations may have been tedious, the group retained a sense of humor. Flanders created a comic ballet titled *Sacre du Mesa* that satirized the culture of Los Alamos; he also danced the part of General Groves. The main prop, Bernice Brode recalled, "was a mechanical brain with flashing lights and noisy bangs and sputters, which did consistently wrong calculations, for example, $2 + 2 = 5$" (Bennis 1997, p. 186).

Most of the wives who became "computers" at Los Alamos, though very proud of having helped with the construction of the atomic bomb, downplay their contributions. They were very glad to support their husbands and to have some idea of what went on in the laboratories, but their main concern seems to have been the survival of their families in the primitive military-camp environment of Los Alamos.

Naomi Livesay was recruited to Los Alamos as a professional with experience in using IBM electric calculating machines. She received a B.A. in mathematics from Cornell College in Iowa and was given an appointment as a research assistant in the Department of Education at the University of Wisconsin. A male counterpart from the same school received a teaching assistantship in the Mathematics Department. He carried only mathematics courses, but Livesay was required to carry education courses in addition to her graduate math courses and twenty hours of research work each week. In the second semester of her first year, Rudolph Langer, one of the mathematics professors, told her, "There is no place in higher mathematics for any woman, however brilliant." A Ph.D. in mathematics, he said, would do her no good; universities clearly did not want her, and the degree would make her overqualified to teach high school (French 1990).

In 1939, Livesay received an education-oriented degree—a Ph.M. in mathematics—from the University of Wisconsin and entered the Depression-era job market. She discovered that the one available job in mathematics near Madison offered male candidates $1,000 a year and female candidates $900. As her summer job in education ran out, she had no job prospects at all. Joe Hirschfelder, who was then a young chemistry professor at Wisconsin and who served as a big brother to the graduate students, suggested that the university's Education Department had not helped her find a job so it would be able to keep her as a research assistant. Fortunately, Hirschfelder had a friend at Princeton Surveys in Princeton University's School of Public and International Affairs who needed a mathematician to do statistics on the costs of state and local governments, and Livesay got the job.

The Princeton Surveys sent her to Philadelphia for a training course in operating and programming IBM electric calculating machines. The machines could read punched cards and do very simple repetitive mathematics. One machine could add or subtract. A second could multiply. Each machine was the size of a large desk, and only the adder could print out results. Some of the machines could be programmed with a plug board. Livesay was given an assistant and spent the next six months designing and implementing plug-board programs, punching cards, and carrying cards from one machine to another.

Work on the surveys quickly proved boring for Livesay, and she was delighted to receive a prestigious Rockefeller Foundation Fellowship to work on a study of education at the University of Chicago. Only when she announced her resignation did her superiors at Princeton suggest that she might take graduate courses in mathematics at the university.

At Chicago she was assigned to work on a psychology project with George Sheviakov, head of guidance work for the Chicago Laboratory School, and Bruno Bettelheim, a refugee with a Ph.D. in art history and a graduate of Freud's school in Vienna. She offered to take courses in psychology to better qualify her for the work, but this pair of remarkable tutors suggested she stay with them, as they could teach her more than the regular faculty. When the Rockefeller Fellowship expired, the director of the psychology project decided he wanted her directly on his staff and arranged for Rockefeller not to renew her fellowship. She refused the staff job.

In the fall of 1940, Livesay found a position as a teaching assistant at the University of Illinois. Women were hired because many male faculty members were being pulled into war-related research. At the time, Illinois had two female full professors and one assistant professor on its mathematics staff. Livesay was a full-time instructor in the fall of 1941. She was not offered a permanent job, but the environment was pleasant and the work was fun. She was seeking more permanent employment when Joe Hirschfelder wrote to offer her a job on a "war project." The letter came in mid-fall of 1943. She went home to her family in Illinois before Christmas and headed for Los Alamos in February 1944, when her security clearance came through.

She arrived at Los Alamos late one afternoon after a long train trip from Illinois. Hirschfelder was at a seminar, but his secretary got Livesay some dinner, and the new employee spent her first night at Los Alamos attending a party. The next morning she learned about the atomic bomb. Hirschfelder's group had been working on the gun model for plutonium, and their project had just been terminated, but a new group was being formed to do calculations related to the implosion weapon. Since the group would be using IBM machines, Livesay was made to order for it.

Hirschfelder introduced her to Hans Bethe, director of the Theory Division, and left her to be interviewed by Stanley Frankel and Eldred Nelson, who would be in charge of the project and wanted her to supervise the operation of the IBM machines that had been ordered. The door was open, the walls were plasterboard, and her interview could be heard out in the hall. To Livesay's surprise, Mary Frankel, Stanley's wife, sat in on the interview, and a very odd character—a dark-haired man about twenty-five years old—kept drifting in and out, making critical comments to Frankel. Finally he settled down in the back of the room and listened to the rest of the interview.

By this time Livesay had real qualms about taking a job in this peculiar environment. She voiced her doubts, and was left alone with the odd male character. He introduced himself as Richard Feynman and urged her to take the job for the sake of the war effort. She accepted his logic and became an assistant scientist before lunch on her first day in Los Alamos.

The IBM machines arrived two or three weeks later, and Livesay went to work supervising a crew of GIs and civilians who kept the machines running twenty-four hours a day. When the shock wave the group was modeling reached an interface between two materials—a point that the machines had not been programmed for—the calculations were carried forward by hand until the wave reached a uniform material again. Livesay and Tony Skyrme of the British mission would do the hand calculations and compare their results, then take the figures back to the machines to continue the main calculation. A single interface calculation took six to eight hours.

The primary problem facing the group was locating errors made by the machines. The IBM calculating machines relied on electric relays, which stuck when dust got into them. The machine room faced an unpaved road that carried heavy truck traffic past open windows. The entire group, including Hans Bethe, the division head, constantly checked the results for errors. An IBM repairman was on call 24 hours a day—because he had been drafted and ordered to Los Alamos.

As the Trinity test approached, the Calculations Group felt increasing pressure to complete the implosion calculations. The tension exacerbated bad relations between Frankel and the staff, particularly the GIs. The soldiers already earned far less than the civilians and had to submit to military discipline as well. Frankel had an unfortunate tendency to make derogatory comments to the soldiers and to lose his temper. He also brought his wife into the machine room, and Mary Frankel assumed authority over the calculating work. Livesay protested to Bethe, with the result that Feynman and Nicholas Metropolis, a physicist from the University of Chicago who later became a pioneer in the development of computers, were placed in charge of the project. The machine operators didn't understand Feynman, but they liked him and worked well for him (French 1990).

The new IBM punched-card machines were used for calculations to simulate an implosion. The implosion simulation required the integration of a hyperbolic partial differential equation in one space and one time variable. The integration was carried out by sending batches of punched cards through a series of unconnected IBM machines. On each run, the machines collaboratively calculated a single integration step; then a set of values was printed and a new batch of cards was punched with the data for the next step.

Each card required one to five seconds of processing by each machine, and a human operator had to intervene at many points (Aspray 1990). Hand calculations were used to check the IBM equipment as it was being installed and made ready. Richard Feynman decided that it would be instructive to organize a contest to determine whether the desk calculators or the punched-card equipment would be more effective at the implosion simulation, so he and Metropolis organized a race. "We set up a room with girls in it," Feynman said. "Each one had a Marchant. But one was the multiplier, another was the adder, and this one cubed, and all she did was cube this number and send it to the next one." The desk calculators kept up with the IBMs for one day. "The only difference was that the IBM machines didn't get tired and could work three shifts. But the girls got tired after awhile" (R. Seidel 1993).

After the contest, the IBM equipment was used for all large calculations, and they went faster. Feynman had discussed the project with John von Neumann, a consultant on hydrodynamics and mathematics, and had decided that the interfaces could be ignored, since the equations rapidly settled down after crossing them. The group received a new "triple multiplier" machine. Finally they worked mainly on the shock wave and the less complex early stages of the blast. Livesay graphed the results of every calculation, and continued to organize the operators with the help of a colleague named Eleanor Ewing, whom she had been allowed to recruit (French 1990).

After Trinity, the Theory Group felt far less pressure and continued work on blast waves from the explosion at a more leisurely pace. Until the Japanese surrender in August 1945, however, the pressure for computing at Los Alamos remained intense (Aspray 1990). The group's results were later used to check the operations of ENIAC, the first general-purpose, all-electronic computer, which during the war was on the drawing board at the University of Pennsylvania. Livesay had instructed von Neumann in the operation of the IBM machines, and he had used the information in designing ENIAC (French 1990).

As pressure to complete the implosion calculations built, Livesay was authorized to recruit Eleanor Ewing, whom she had met at the University of Illinois. Ewing held a B.A. and an M.A. in mathematics from the University of Illinois. As part of her program, she had enrolled in physics. When she walked into the classroom with a male acquaintance, the professor called out, "Miss Ewing, we do not mix sexes in physics. We have the girls' row down in front, please." Not surprisingly, she was the only woman in the class. Despite her ability in mathematics and physics, the professor made her rather slow lab partner team leader. Nevertheless, Ewing received A's in physics and in all her math classes, became a junior Phi Beta Kappa, and was a teaching assistant in mathematics (Ehrlich 1991).

In 1943, Pratt and Whitney Aircraft in Hartford, Connecticut, hired Ewing as a new M.A. to teach elementary mathematics. Her class consisted of women who were interested in becoming engineering aides to help the war. The job was not terribly exciting, but it was considered "essential to the war effort"—phrasing that meant that she could leave only with the permission of her supervisor and the War Manpower Commission. In August 1944, a call from Livesay invited her to an unknown project in an unknown place in New Mexico. She took the offer.

A letter marked "Confidential" arrived from the War Office. It contained a train ticket from Hartford to Chicago and instructions to ask for another letter from the Western Union office in Chicago's Union Station at 10 P.M. on the day of her arrival. She had to show her second letter to military policemen before being allowed to board the train to New Mexico. Ewing found herself the only woman on a troop train for a two-day trip. She left the train at Lamy, a wide spot in the road that served as the station for Santa Fe. Livesay met her and escorted her to Los Alamos.

Her introduction to the Manhattan Project came from Stanley Frankel, the leader of the calculations project, who sat cross-legged on a table and announced that Ewing had come to the mesa to make an atomic bomb.

Her immediate impression was that she might have to undertake the project alone. She couldn't even remember what an atom was.

Ewing and Livesay ate their meals at the Fuller Lodge, where they met such notables as Niels Bohr. Their white badges admitted them to the weekly colloquiums, which, she wrote in a letter, "gave us an opportunity to hear Oppenheimer argue with other project leaders." Ewing was assigned to the Calculations Group and worked as Livesay's assistant in keeping the IBM machines running. The two young women shared an office, which was used for meetings by the group when John von Neumann came to Los Alamos. Von Neumann had excellent manners and rose whenever a woman entered the office. More important, he gave Livesay and Ewing mathematics lectures. If they used the office blackboard, it had to be erased afterward for security reasons (Ehrlich 1991).

The success of the IBM equipment led to interest in the more advanced computers that were then being developed. Some calculations from Los Alamos were performed on ENIAC. ENIAC, which had been under development since June 1943, was designed at the University of Pennsylvania for the Ballistics Research Laboratory at the Army's Aberdeen Proving Ground. John von Neumann's wife, Klara "Klari" von Neumann, became one of the first computer programmers. She, her husband, and Nicholas Metropolis applied ENIAC to Monte Carlo techniques in February 1948. Also after the war, the von Neumanns and the physicists Cerda and Foster Evans worked on computations on the newly built electronic computing machines (Ulam 1976).

Edward Teller's wife, Mici Teller, was engaged in hand calculations using desk calculators early in the war effort. After the war she collaborated with colleagues in calculations on Los Alamos's MANIAC computer, and with her husband and several other scientists was a coauthor of the first paper on Monte Carlo calculations. This paper developed a general modified Monte Carlo method suitable for fast computing machines for investigating such properties as equations of state.

## Conclusions

Perhaps no technical area involved in the Manhattan Project has changed more than applied mathematics. Modern computers easily handle the thousands of individual computations that proved such a challenge during the war. The extremely complex hydrodynamic models that supported the design of the plutonium implosion bomb were a technical tour de force. Female mathematicians and calculators played a major part in organizing and carrying out this massive effort. In the process, several of them gained experience that led them to prominent roles in the rapid development of computers after the war.

Calutron operators in the Y-12 Control Room. *(Department of Energy photograph by James E. Westcott, Neg. #Y-12 142)*

The "Gunk" extraction towers where enriched material from the Y-12 calutrons was collected. *(National Archives, RG 434-OR, Box 23, Notebook 69, Y12-138)*

Sample wash line at the Y-12 plant. *(National Archives, RG 434-OR, Box 23, Notebook 69, Y12-138)*

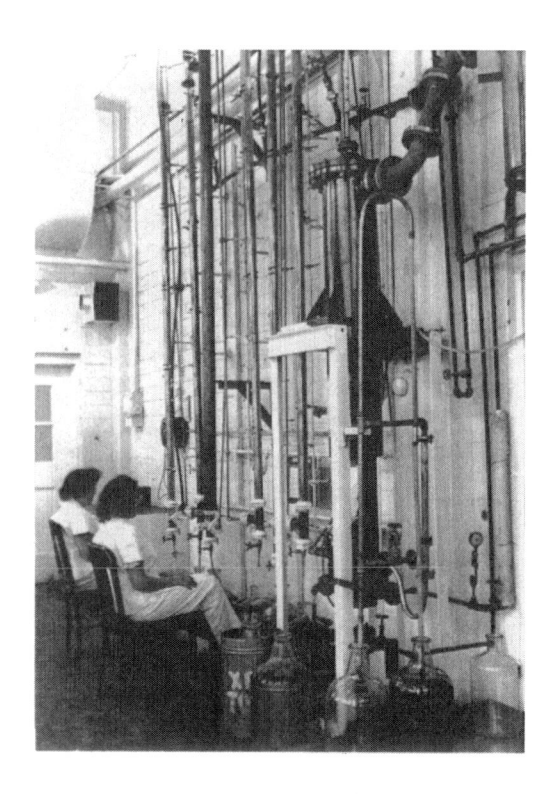

Process equipment at the Y-12 plant. *(National Archives, RG 434-OR, Box 23, Notebook 69, Y12-144)*

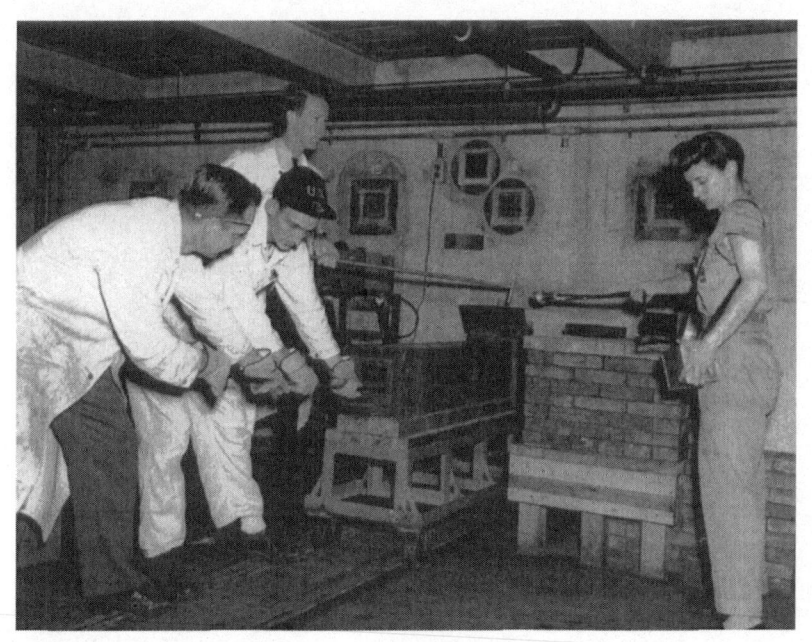

Unloading isotopes from the X-10 plant. *(National Archives, RG 434-OR, Box 22, Notebook 65, MED 308)*

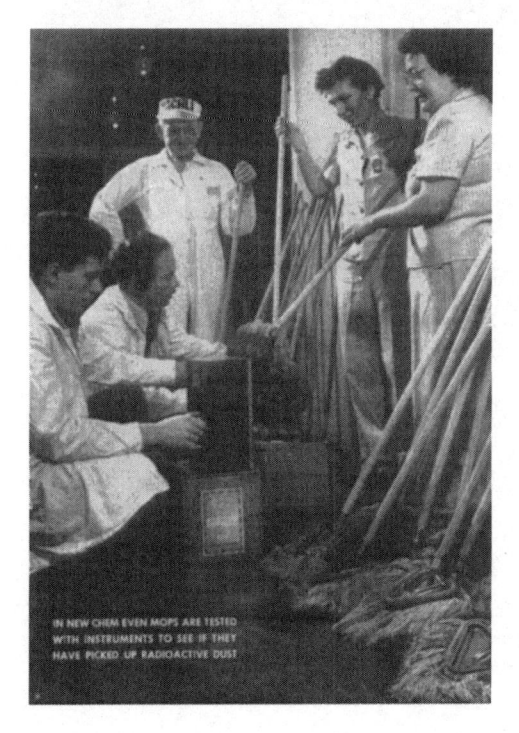

Radiation safety complicated life at the Manhattan Project sites. Here mops are checked for contamination in the Met Lab. *(Photograph by Fritz Goro of Life magazine; used courtesy of the Lawrence Berkeley National Laboratory)*

Eleanor Ewing Ehrlich poses beside Jumbo, the container built to surround the Trinity test and recover plutonium in case of a fissile. It was never used. *(Eleanor Ewing Ehrlich, private collection)*

The B reactor at Hanford under construction in September 1944. The untested design of this massive construction project demonstrates the rush from laboratory to full-scale production. *(National Archives, RG 434-OR, Box 14, Notebook 41, MED 382)*

# 6  Biologists and Medical Scientists

WHEN IT BECAME clear, early in the course of the Manhattan Project, that workers would be dealing with new substances and unprecedented amounts of radioactivity, the organizers instituted biological and medical programs. The biomedical concerns were threefold and had little to do with intellectual curiosity. They reflected the wartime urgency of the project. In the words of the Department of Energy's Web site, they were:

1. If enough knowledgeable personnel were disabled or killed from exposures, maintaining the intended schedule for production of atom bombs would not be possible.
2. If enough knowledgeable personnel were disabled or killed from exposures, keeping the project secret would be difficult.
3. If excessive radioactivity spread from laboratories and production facilities, it might be detected through adverse health effects. This detection would likewise threaten to compromise secrecy (DOE Roadmap).

However pragmatic their reasons, the leaders of the Manhattan Project were forced to establish research programs in radiation biology and to pay close attention to the health of the workers.

To a limited extent, scientists had recognized and studied the biological effects of radiation almost from the discovery of X-rays. Before World War II, however, the most hazardous radioactive material known was radium. Its industrial use in producing luminous watch dials led to high rates of cancer and bone disease in workers. Yet the total amount of radium in the world before the war was no more than 100 grams (Goodman 1996).

When the war began, Edith Hinkley Quimby already had an international reputation for her studies of the physics of radiation exposure, and she made her experience and capabilities available to the Manhattan Project. Hinkley was born in 1891 and grew up in the Midwest. A high-school teacher stimulated her interest in chemistry and physics and taught her to answer some of her own questions by conducting experiments in the laboratory. She decided to major in physics and mathematics in college, and was awarded a full-tuition scholarship to Whitman College in Walla Walla, Washington. The math-physics major had never before been taken by a female student at Whitman (Kass-Simon et al. 1990; Hacker 1987; Rossi 1982).

Like many other female scientists of her generation, Hinkley taught at the high-school level before going on to graduate school. She taught high-school chemistry and physics for two years before obtaining a teaching fellowship at the University of California at Berkeley. She earned her master's in physics in 1916. At the end of her first year of graduate work, she married a fellow physics student, Shirley L. Quimby (Yost 1960; McMurray et al. 1995; Bailey 1994).

In 1919, Edith needed to find a job in New York in order to support her husband, who had become a junior faculty member at Columbia. Because she was married and had only a master's degree, the only job she could find was as an assistant to the radiologist at the New York Memorial Hospital for Cancer and Allied Diseases. Even getting this modest position was somewhat of a coup, since not one other woman in America was engaged in medical physics research at the time (Yost 1960; Rossiter 1982).

At Memorial Hospital, Quimby worked with Gioacchino Failla in developing the use of radioactive therapy for cancer. The field of radiological physics did not exist when Failla and Quimby began their work, even though radiotherapy was being practiced in a number of hospitals. Dose rates were still measured in erythema doses: the amount of radiation exposure needed to burn the skin enough to redden it. In the 1920s Quimby introduced the first film badges for monitoring exposure to radiation, a development of considerable practical importance. In a 1984 interview, she said:

> Well, this was fairly early on. I think it was '23 or so, somewhere back there, in line with our appreciation of the dangers the technicians might have. We knew that the radiation would blacken photographic plates or films. What I did there was to take a big X-ray film and cut it up into strips, and then wrap each strip in black paper. Then we could put them around and see how they got dark. But then you had to have a standard, just like the film-badge people do now. So I would take a series of films and put the radium applicator over it for certain periods of time, such that the exposure to the film was in erythema doses. It was millicurie hours at a

given distance. And so we got films—well, the one that shows up pretty good in the middle was about 25 thousandths of an erythema dose. You see, the film is so darn sensitive. But we could make a standard set, and then we could have films all around the lab or on people and see how much radiation they were getting. So this, I think, was the first film-badge service that was ever developed. (Quimby 1984)

Quimby collaborated with several physicians in investigating the medical effects of radiation, and was the author of some especially important work on radiation effects on the skin (Rossi 1982). When Marie Curie came to New York in 1921, she visited Quimby and Failla at the hospital (Quimby 1984).

Quimby's lectures at Memorial Hospital developed into a book that became the bible of radiologists during World War II. The fifty papers she published before 1940 set practical dose specifications for radiotherapy, and she was awarded the Janeway Medal of the American Radium Society in 1940. In 1941 she became the first woman since Marie Curie to win the gold medal of the Radiological Society of America (Rossiter 1982). She moved to Cornell University in 1941, and in 1943 to the Columbia College of Physicians and Surgeons, where she established the Radiological Research Laboratory (Rossi 1982).

Quimby worked part-time at Columbia, studying the medical effects of radiation exposure (Rossiter 1995). Her work with Failla involving total-body irradiation was of particular interest to the Manhattan Project (Friedell 1994). Working at Columbia and Oak Ridge during the war, Quimby expanded her activities to include the use of radioactive nuclides in medicine. She made very important contributions to the complex subject of dose calculations for emitters absorbed by the body, and published papers on the dosimetry of internally administered radioactive substances (Rossi 1982; Quimby 1984; OHRE 1996).

## Biology Begins in the Manhattan Project

Early in 1942, biomedical researchers began experiments to establish exposure standards for the workers in plutonium production and isotope separation. In 1943, the Manhattan Engineer District designated a chief of the Manhattan Project Medical Section. The Medical Section was responsible for industrial safety, medical care, and funding and coordinating biomedical research at a group of associated universities.

Until the war, biomedical researchers at these universities had used radiation as a tool to study biological systems and disease. The Manhattan Project shifted their emphasis to study the metabolism of radioactive materials and health effects of radiation (OHRE 1996).

Biomedical recruits to the Manhattan Project often had little idea of the field they were entering. For example, Miriam Posner Finkel, a Chicago native, had received her Ph.D. in zoology from the University of Chicago in 1944. She had just married Asher Joseph Finkel, a fellow student, and she needed a job (*American Men and Women of Science*). She followed up a friend's rumor that the Metallurgical Laboratory on campus was hiring biologists. At her interview, almost nothing was said about the job except that it did involve biology. The interviewer said he would initiate security proceedings, and three days later she started work (Finkel 1976).

Finkel worked as an associate biologist at Site B—"the brewery"—on establishing the basic toxicity levels of radionuclides. She and her colleagues took over all short- and long-term studies of the toxic effects of radiation on animals that had been injected or otherwise treated with samples of the radionuclides, including fission products and transuranic elements. In particular, they used some of the first samples of "product," as plutonium was then called (*American Men and Women of Science*; Ehret 1992; Kisieleski 1993; Lisco et al. 1947). She reported on studies of the chronic toxicity of radiostrontium (OHRE: Health Div. report 1945), and she published extensively on fission products and transuranics. After the war she became a senior biologist at Argonne National Laboratory.

In biomedical studies, as elsewhere in the Manhattan Project, secrecy prevailed. "We didn't want anybody to know that we were working on radiation," says one worker. "Work on animals was secret, because we didn't want anyone to be aware of this. As a matter of fact, publications in the literature, once they came into our office, were stamped 'secret'" (Friedell 1994, p. 17).

In December 1942, when Enrico Fermi and his colleagues achieved the first self-sustaining nuclear chain reaction, "the problems of the Health Division," as one of the leaders of the division put it, ceased to be "mainly theoretical" (Hacker 1987, p. 33).

Initially the Health Division contained two sections: clinical and experimental. The Clinical Section established standards and procedures to protect the health of laboratory workers and the public. The

Experimental Section investigated the effects of ionizing radiation to support the clinical efforts and to gain an understanding of the biological effects. Biological researchers investigated both cellular exposure and the effects of exposure on live animals. Eventually the division assumed responsibility for training personnel to take over the labs that were to be built at Clinton and Hanford.

The Manhattan Project needed clinicians. Dr. Margaret J. Nickson and her husband, Dr. James J. Nickson, were among the physicians hired by the Health Group. Margaret was born in Vermilion, Ohio, in 1914, and received an A.B. from Transylvania College in 1936. She married in 1939, and received an M.D. from Johns Hopkins University in 1940, as did her husband. After an internship at a hospital in Pennsylvania, she joined the Metallurgical Laboratory and worked as a research associate from 1943 to 1946 (*American Men and Women of Science*; Hacker 1987; Kathren et al. 1994; G. Sacher 1952).

Nickson undertook the blood testing of Manhattan Project workers and examined the changes that occur in the fingers of people occupationally exposed to radiation. She studied the effect of single-dose X-rays on the nail fold area of human subjects and investigated atrophic fissures and ridge flattening in finger impressions (OHRE, Factsheet Human Experiment 101 1947; OHRE, Harvey letter, U. of Rochester Strong Memorial Hospital letter Oct. 23, 1945; C. L. Marshall memo 1947). Drs. Margaret and Jim Nickson were among the early staff members who stayed at Site B, and subsequently both worked at Argonne National Laboratory (*American Men and Women of Science*).

Melba Robson, Edna K. Marks, and Evelyn Gaston provided technical assistance in extensive work conducted by James Nickson on blood changes in humans after total-body irradiation; the results were presented in *Metallurgical Laboratory Report CH-3868* and subsequently published in the open literature (Nickson 1951). Melba Johnston Robson was employed by the University of Chicago in the mid-1930s as a laboratory technician/medical technologist (D. Robson 1995). She conducted studies of blood chemistry and taught techniques to laboratory technicians (Watson 1995). She ran complete blood tests on rats and rabbits that were placed at varying distances from CP-1—the reactor under the west stands of Stagg Field.

Biology at the Met Lab initially occupied the Knickerbocker stables at Site B, the old brewery a few blocks south of the quadrangles of the main campus. When the researchers moved in, there were reportedly still "faint evidences" of the previous occupants. Site B was overcrowded from the beginning by the growing biomedical section and a machine shop, and in spite of many expansions remained overcrowded to the end of World War II.

The group's piecemeal expansion led to a corridor layout that was mystifying to visitors and new employees. The biologists worried constantly about air-conditioning failure. When this happened, only immediate action would save the experimental animals. Site B was cooled by a jury-rigged setup of air conditioners, and the site had a distinctive odor compounded of the effluvia of animal colonies and muffle furnaces (G. Sacher 1952).

Most of the workers in the stables were very young, and many of them left little record of their activity on the Manhattan Project. Elaine Katz Bernstein was born in Baltimore in 1922. She received a B.A. from Goucher College in 1941 and an M.A. from Oberlin College in 1942. As a junior biologist at the Met Lab from 1943 to 1946, she participated in studies of the hemolytic effects of radiation. A report on this work was subsequently declassified (Schwartz et al. 1946). A biochemist and later a science writer and editor, Bernstein conducted research at Argonne National Laboratory in clinical chemistry, liver function, toxicology, radiation-induced biochemical changes, and leukemia (*American Men and Women of Science*).

Marcia White Rosenthal grew up in Chicago and received an A.B. from Cornell University in 1943. As a biochemist at the Met Lab, she generated the early data on plutonium disposition in chelation therapy. She later became a vertebrate zoologist, receiving her Ph.D. in zoology from the University of Chicago in 1949. Her areas of specialization were embryonic sex differentiation, metabolism of radiocolloids, therapy of experimental metal poisonings, carcinogenicity of plutonium, and chelation in drug action (*American Men and Women of Science*).

Janet Tasseneau studied the effects of a high-intensity dose of radiation on a part of an animal—a foot, for instance—and compared them with the results of equivalent doses given more gradually to entire animals (Kisieleski 1996). Dorothy Wallace monitored workers, buildings, and equipment for beta and gamma radiation (OHRE: Health Div. report 1945). After the war, she worked in radiological physics at Argonne National Laboratory. Lillie Mae Porter worked first as a laboratory assistant in plutonium chemistry and then as a laboratory technician in the Health Division. She was coauthor of a Met Lab report on the hemolytic effect of radiation (Schwartz et al. 1946).

Other women came to the Met Lab Biology Division with considerably more seniority. Born in New Freedom, Pennsylvania, in 1909, Elizabeth Painter received an A.B. from Goucher College in 1930 and a Ph.D. in physiology from the University of Maryland in 1938. She taught physiology and chemistry at the University of Maryland School of Medicine. In 1937 she moved to the College of Physicians and Surgeons at Columbia University.

The Met Lab recruited Painter as a research associate in biology in 1943, and she worked there until 1945, with time out in 1944 to serve as a civilian with the Office of Scientific Research and Development. Painter, an inhalation toxicologist, studied the toxicology of fission products in laboratory animals. She also experimented with the physiological effects of ionizing radiation in animals, including effects of single doses of X-radiation on dogs (*American Men and Women of Science*; Kisieleski 1996; Prosser et al. 1947; OHRE: Health Div. report 1945).

The workers in the crowded biology laboratories hurried through their studies of the effects of transuranic and fission products on animals. The Hanford reactors were under construction, and the project leaders desperately needed standards that would let them know how to protect the workers there. The pressure to finish the biological experiments was nearly as intense as that on the scientists developing the isotope-separation plants and the production reactors.

Katherine L. Lathrop was born in Lawton, Oklahoma, a frontier Army post, in 1915. Her maternal grandparents had moved from Iowa to what was in 1902 still Indian Territory; her paternal grandparents had moved to the Oklahoma Territory from Texas a few years earlier. Her parents lived within a few miles of each other growing up. She was the first child in the family (OHRE 1996).

Lathrop attended Oklahoma State University, where she received a B.S. in biology in 1936, a B.S. in physics in 1939, and an M.S. in chemistry in 1939. She married while doing her graduate work, and in 1939 moved with her husband to New Mexico, his home state. In early 1942, the couple moved to Laramie, Wyoming, where Lathrop became a research assistant at the University of Wyoming. She worked in the poisonous-plant laboratory, filling in for workers drafted for military service, while her husband taught chemistry to enlisted personnel.

In 1944, Lathrop's husband was admitted to the medical school at Northwestern University, and they moved to Chicago. "We thought that we had enough money from the sale of our house and savings," Lathrop says. "However, living in Chicago cost much more than living in Laramie." They needed money badly. Fortunately, one of her husband's friends told them about a secret project on which he had worked before entering medical school. The project, he said, had jobs for people trained in scientific fields. Lathrop applied by mail and was asked to appear for a personal interview. The person who interviewed her thought that she would fit well in either the Chemistry or the Biology Section of the Manhattan Project. The first person he called—in the Chemistry Division—wasn't there, but his contact in the Biology Division was, and Lathrop went at once for an interview. She was hired the same day.

That unanswered phone call, Lathrop later said, was decisive in her career. She might otherwise have gone into straight chemistry, but her experience in the Manhattan Project, where she studied the biological effects of fission products, was valuable preparation for the contributions that she later made to the development of nuclear medicine.

Lathrop joined the Manhattan Project as a junior chemist. She remembers her first day of work:

> Dave Anthony said, "I think it would be best for you to read for a few weeks in order to learn something about the work we are doing."
> He took me upstairs to a room, maybe 15 by 30 feet, lined with bookshelves holding about six books. There was a door like those used on the safety deposit room in a bank that, I later learned, was a depository for classified documents from the various sites working on the bomb project. An attendant sat at a desk to check out documents requested from the "vault." (OHRE 1996)

Lathrop used this opportunity to acquaint herself with some of the work that had been done on radioactivity.

> Of course, I knew very little about the various types of radiation, but the work I saw going on around me seemed mostly to involve handling animals. In high school I had chosen Chemistry and Physics for science because I wanted nothing to do with the cats being dissected in Biology. Faced with earning enough money to care for my two children and get my husband through medical school, I quickly decided that I could learn

to work with animals. I had taken a nonlaboratory Anatomy course in college. Dave told me that we were working on some radioactive explosive material that might win the war for the Allies. He solemnly asked me to tell no one what I was working on, not even my family. I never did until after the bomb was dropped. (OHRE 1996)

Lathrop was not put off by the military presence in the project. "Some enlisted people in Army uniforms were working along with civilians, in the laboratories, doing the same work," she said. "Having grown up next door to Fort Sill, where I daily saw, on the streets of Lawton, men of all ranks in uniform, the military was just part of the background, and did not impress me."

One of the few African-American women involved in the technical work of the Manhattan Project, Mildred M. Summers was born in Vicksburg, Mississippi, in 1913. She attended Natchez College from 1929 to 1931, and then, like many black people at the time, came to Chicago seeking work. From 1932 to 1937, Summers worked as a typist at Chicago Hospital. By early 1942 she was working in a hospital on Michigan Avenue, but in the fall she quit the job. A friend told her to apply for a job at the Ryerson Laboratory building on the campus of the University of Chicago. She visited the lab, where Glenn Seaborg, head of the Chemistry Section, interviewed her. In February 1943 she received a telegram telling her to report to work as a laboratory helper at the New Chemistry building (D. Sacher 1995; Kathren et al. 1994).

Summers' first job was to count samples; she recalled that Art Jaffe, a chemist, taught her how to use alpha and gamma counters. Later she also pipetted and decontaminated the glassware. "The samples would be pipetted out in our lab," she said, "and we would take them across the street to the West Stands to be counted" (D. Sacher 1995).

As a matter of course, Summers and her coworkers had their blood counts measured every other week. On one occasion she became contaminated after pipetting a radioactive sample carelessly, and had a blood count conducted by a woman involved in biomedical studies at Drexel House (D. Sacher 1995).

In May 1944, after working in groups concerned with the bismuth-phosphate separation process and procurement and technical services, she rose to the level of a technician in the Extraction-Decontamination Group (Kathren et al. 1994). Records show that she was scheduled to leave the group on June 30, 1945 (Kathren et al. 1994). In her oral memoirs, however, Summers says that in April 1945 she transferred to the Biology Division and worked at Site B. She studied the effects of drugs and radiation on mice before leaving the project. Summers continued work at the "new site in the country"—Argonne, that is—and after 1952 she worked at Argonne National Laboratory with Agnes Stroud, a Native American scientist who had worked at Los Alamos during the war (D. Sacher 1995).

## BIOLOGY MOVES TO CLINTON

As the production plants and test reactor began operation, the need arose for a biomedical program at the Clinton Engineer Works, so a branch of the Met Lab's Biology Division moved to Oak Ridge in 1943. Some members of the Chicago staff made the move, and new workers joined them at Clinton. The startling switch from urban Chicago to the muddy streets of the Tennessee boomtown was not the only change the Chicago workers had to adapt to: At Clinton, the focus of their work was protecting and treating human beings rather than studying animals.

Melba Robson moved to Clinton with the Chicago group and did general medical technologist's work, such as blood tests and hospital support, until late 1945 or early 1946. She was well aware of the overriding purpose of her work. Her son remembers:

> At the time of the first test at Los Alamos, Melba was on vacation in northern Minnesota, where her family lived. She was staying at her sister's wilderness fishing camp. She knew that the test was planned to occur while she was there.
> One of the concerns Melba had heard about regarding the test was the possibility that it would ignite a self-sustaining fusion reaction on Earth. Understand that at that time fusion was poorly understood. The only sure knowledge was that the ignition conditions were in the range of an atomic explosion. Earth's surface is rich in elements light enough to possibly fuse exothermically, and the water vapor in the atmosphere carries substantial amounts of hydrogen. Some people thought there was a significant chance that the Los Alamos blast would ignite a fusion fire that would vaporize the Earth's surface, converting the hydrogen in the seas to helium and leaving a lunar landscape behind. Although she gave the concern little credence, the vacation still had an atmosphere of terror that was only relieved by the news that the test had been successful.
> The secrecy required added to her problems. She could not explain to her family why she would immediately leap on every newspaper deliv-

ered and read it from cover to cover, looking for a hint that the explosion had already occurred. (A. Robson 1995)

Because the Manhattan Project oversaw entire communities at Clinton, Hanford, and Los Alamos, the medical section also ran clinical programs to treat routine health problems. Only a fraction of the many Medical Section staff members were engaged in research.

The first physician at Oak Ridge was Gladys Morgan Happer. She was recruited by her brother, Karl Ziegler Morgan, a physicist. Arthur Compton had sent Morgan, who was subsequently widely regarded as one of the founders of the field of health physics, from the Met Lab to Oak Ridge to organize a program aimed at protecting workers from large amounts of ionizing radiation. He became the head of health physics at Clinton Engineer Works (Goodwin 1991).

In the words of Happer's son William, "Karl, a good physicist but with no training in biology, keenly felt the need for a physician to whom he could talk, and fate presented his sister, my mother, a medical missionary who had been sent back to the United States from India at the outbreak of World War II. My mother had obtained an MS in organic chemistry from the University of North Carolina before going to medical school at Women's Medical College in Philadelphia, so she was able to talk to physicists and chemists at Oak Ridge more or less on their own terms" (Happer 1991).

Morgan described his sister as "the smartest person I have ever known." She had, he said, "a tremendous memory" and "most of the brains in the family" (Morgan 1991). She completed her master's degree in chemistry in one year, then entered the University of North Carolina medical school. She was at the top of the class, but the school didn't want to recommend her to a major medical school because she was a woman, so she went to Women's Medical College. After graduating, she traveled to Paris, where she learned French and conducted some research before leaving for Geneva to study languages and conduct further research in medical school.

Gladys Morgan felt obligated to the Lutheran Church, which had paid for her medical school, so she went to India as a medical missionary to take charge of a hospital (Morgan 1991). While in India she married William Happer, a Scottish physician who was serving as a lieutenant colonel in the Indian Army. The couple's first two children were born in India, the third in the United States, to which Gladys had returned when her husband's unit left India at the beginning of the war. She stayed with her parents until she moved to Oak Ridge (Morgan 1991).

Then as now, child care presented problems for working women. William Happer recalls:

> I was a small child, about three or four years of age at the time we lived in Oak Ridge, and I remember the guard towers and barbed wire, the board walks, the muddy streets, and the "temporary" buildings, some of which may still be there. Since my father, Col. William Happer, a Scottish physician in the Indian Army, was trying to cope with Germans in Iraq, and later in North Africa, my Mother and I were alone in Oak Ridge, and my Mother had quite a bit of trouble finding someone to look after me while she was at work, which was quite a few miles out of town at X-10, now the Oak Ridge National Laboratory. I remember that one of my babysitters was the wife of a scientific glassblower at the laboratories. (Happer 1991)

Gladys Happer arrived in Oak Ridge in the fall of 1944. "A number of employees came and said how happy they were to have her," her brother recalls. "She was the best doctor they had ever had."

She participated actively in early clinical studies of the effects of radiation. "At Oak Ridge, my mother had to cope with serious cases of radioactive contamination," William Happer writes. "I remember one night that my mother came home very late after having shampooed the plutonium-contaminated head of one poor worker again and again to remove the stubborn radioactivity. Uncle Karl ended the day with the helpful suggestion that she cut the man's head off."

Not all of Happer's problems were medical. According to her son, the "military mind" was a constant source of friction:

> For example, the gravel road from the town of Oak Ridge to the laboratories produced a huge cloud of dust in the mornings and in the evenings when the buses and cars took people to and from work. This caused serious respiratory problems for many workers and endless headaches for my mother who had to treat them. She tried again and again to have the road blacktopped, but the military commanders refused because "it would draw attention to the operations." You could see the cloud of dust all the way to Knoxville, and everyone knew that there were lots of people in Oak Ridge. My mother was never able to make sense of this one. (Happer 1991)

Happer stayed at Oak Ridge throughout the war, then returned to India with her son to rejoin her husband. "At the end of the war, we went back to India to meet my father for a few short years of togetherness before the end of the British Raj forced us to leave India," recalls William Happer. "Our first stop as former colonials was my father's native Scotland. A year or so later we returned to the United States, where prospects were better for families like ours, whose lives had been changed by war. My mother thought about returning to Oak Ridge, where she had a standing offer from her wartime friends, but she did not want to leave her children, a decision which greatly enriched the lives of me, my brother Ian and my sister Elizabeth."

Gladys Morgan Happer died in a head-on collision with a drunken driver on Thanksgiving Day 1963 (Happer 1991; Morgan 1991). In her brother's words, she had a brilliant career with only one problem: She was a woman. This country, he said, was very slow and very backward in allowing women to get out of the kitchen and into science. Happer's three children clearly benefited from their mother's example, becoming a physicist, a neurosurgeon, and a teacher (Goodwin 1991; Morgan 1991).

Although the Health Division at Clinton nominally cared for workers' health and safety, so little was known about the hazards of radiation that much of division's work was in effect basic research, particularly in developing techniques for monitoring radiation exposure. Mary Rose Ford, a health physicist, became a senior member of Karl Ziegler Morgan's division (Morgan 1991). She did her undergraduate studies at Western Kentucky State University in mathematics and chemistry (Ford 1979), and graduate work at the University of Kentucky and the University of Tennessee. For a number of years she taught school in the public schools in Christian County, Kentucky. When World War II began, she accepted a job working in a medical laboratory at nearby Fort Campbell and became a registered medical technologist. The physician for whom she worked was sent overseas. When he later returned to the Oak Ridge hospital, he recruited Ford to the laboratory there, a job in which she remained until her marriage in 1948.

In that year, Morgan invited Ford to help him in working with internal dosimetry. She was so successful at this that she became known in the lab as "Mrs. Internal Dosimetry" (Ford 1979). She performed some of the early calculations of mass-absorption coefficients for X-ray and gamma-ray photons, using Frieden calculators, as this was before the first computer became available at Oak Ridge. She and a colleague, Mary Jane Cook, studied how quickly nuclear radiation loses energy in materials and contributed to the development of several handbooks. Ford conducted some early studies with phantom models and performed pioneering studies using the Monte Carlo method (Ford 1979).

Other women working at Clinton included Mrs. T. H. Davies, a physician, and Mrs. Russell, a biologist. Most of the medical research staff were women, as were most of the nurses involved in caring for the early victims of radiation exposure. Gladys Lenoff, a nurse in the hospital at Oak Ridge, treated cases of radiation exposure.

## FEMALE BIOLOGISTS AND MEDICAL PERSONNEL AT HANFORD AND LOS ALAMOS

Manhattan Project officials chose Hanford as a site for plutonium-production reactors and the associated separation plants because Clinton lay too close to populated areas. The plutonium-production process would produce large quantities of radioactive materials, use toxic chemicals, and generate toxic and radioactive waste. In accordance with the Manhattan Project's practice of using private industry, E. I. Du Pont de Nemours was selected to construct and operate the facility.

Construction began in 1943, and by the end of 1944 the first Hanford reactor, the B reactor, began operating. The chemical-separation plants needed to extract plutonium from irradiated slugs were finished shortly afterward. Within two years of starting, the huge complex of the Hanford Engineer Works was in place; the site was operationally complete by early 1945. In a few months, Hanford had produced the plutonium for both the Trinity explosion at Alamogordo and the Nagasaki bomb (DOE Roadmap 1996).

Concerned for the safety of its workers, Du Pont established its own clinical and occupational medical programs under the general supervision of the Manhattan Project Medical Section. In addition, Du Pont conducted various kinds of research and development at Hanford, although these were of low priority compared with production. Substantial work went into programs seeking effective ways to monitor people for radiation exposure. The Hanford health physicists developed dosimetry devices such as film badges and methods of interpreting dosimetry measurements. Because the facility released substantial radioactive emissions, the Hanford staff also worked on stack and environmental monitoring programs (DOE Roadmap 1996).

Jane Hall, a physicist, worked on health physics and the development of instrumentation at Hanford. We know that a number of other women were involved in monitoring staff members for exposure, but we know little about these women as individuals. Even though Margaret M. Shaw, the nursing director, was one of two female members of the top management at Hanford, we know nothing of her life. Hanford was so large and its population so transient that we've found it more difficult to learn about the women who worked there than to learn about those who worked at the other project sites.

At Los Alamos, a Health Group reporting to the director was part of the administration from the beginning. In addition to monitoring and record-keeping, the Health Group established and disseminated health standards for safe levels of exposure to radiation and to radioactive and chemical poisons. The group did not figure in the early plans for Los Alamos, but the need for such work became obvious before the laboratory was constructed (OHRE: History of the Los Alamos project, v. 1, draft, 1996).

Like the biology groups at other project sites, the Los Alamos group grew quickly and recruited talented women, such as Anne Perley, who received a B.A. in chemistry from Grinnell College in 1927 and an M.A. in biochemistry from the University of Nebraska college of medicine. She worked for a year in a chemistry laboratory before joining the Department of Pediatrics at Washington University in St. Louis as an instructor in biochemistry. She was recruited to the Health Physics Group at Los Alamos in July 1944. There she monitored workers for exposure to radionuclides, such as plutonium and polonium, and conducted studies on the distribution and excretion of plutonium.

In some ways the work at Los Alamos grew more dangerous after the end of the war, as fissile uranium-235 and plutonium-239 became more available for research. One task of a group testing the purity of the fissile materials was to push the hemispherical nuclear components of future bombs together with a hand tool while measuring the radiation being emitted. The danger was understood; a member of the group, a physicist named Louis Slotin, had already died after a lab accident. During measurements on the criticality of a uranium sphere—the last such test he was scheduled to perform—Slotin was decreasing the separation between hemispherical beryllium reflectors when the tool he was using slipped and the reflector closed completely, causing a runaway chain reaction (Perley 1991; Hoddeson et al. 1993; Groueff 1967). A historian has described the incident:

> With a tinge of nostalgia Slotin performed his critical-assembly test one last time, on May 21, 1946, for the edification of his successor, Al Graves, and six other men who chanced to be in the room. Slotin, who had conducted this test 40 times, approached this final exercise with his usual confidence. The slim, serious scientist pushed the pieces toward one another as usual but suddenly he slipped and the room filled with a blue ionization glow. The Geiger counter chattered, Slotin lunged and flung the hemispheres apart while one of the men yelled, "Let's get the hell out of here!" Seven men charged out of the room, their lives saved by Slotin, who knew instantly that he had taken a fatal dose of radiation. His first concern was Al Graves, who had been at his shoulder. Would he survive? He put the computation to Graves' wife, Elizabeth, without telling her what happened. Mrs. Graves believed herself a stoic; she once dismissed Hiroshima as nothing worse than napalm. But she froze when she learned who the subject of her calculation was. (Rapoport 1971, p. 126–127)

The technique used to learn how much radiation a person had absorbed involved studying the sodium and inorganic phosphorus present in samples of blood and urine. Only after the elements had been concentrated by chemical techniques could the radioactivity induced in them by the high neutron dose be gauged (Hoffman and Hempelmann 1957). At the time of Slotin's accident, only Anne Perley knew how to perform this procedure. Two nurses, Pearl Leach Gordon (Gordon 1991) and Roberta Harvey Jones (R. Jones 1991), were also involved in treating the casualties of this accident.

Slotin was the only person to die as a result of the accident. Al Graves became temporarily bald and developed cataracts and other symptoms of exposure to neutrons, but survived and went on to direct nuclear tests in the Pacific (Fermi 1954; Rapoport 1971).

Anne Perley understood the goal of the Manhattan Project and managed to observe the test of the atomic bomb at Trinity. After the war she conducted biochemical research related to pediatrics.

Los Alamos recruiters searched the Santa Fe area for talent. Agnes Naranjo Stroud, a member of the Tewa tribe of the Santa Clara Indian Pueblo, was one of the few Native American women involved in technical work at Los Alamos during the Manhattan Project. Stroud, born in Albuquerque in 1922 (*American Men and Women of Science*), received a B.S. from the University of New Mexico in 1945 (O'Neill 1979). She worked as a research technician in hematology at Los Alamos from 1945 to 1946. Hematology was a medical field of particular concern during the Manhattan Project because it was known that exposure to radiation caused changes in the blood. Stroud later obtained a Ph.D. and became a staff member at Argonne National Laboratory; eventually she returned to the Los Alamos staff as a cytogeneticist.

Rose Epstein Frisch was born in New York in 1918. She received a B.A. from Smith College in 1939 and an M.A. from Columbia in 1940. She married in 1940 and received her Ph.D. in physiological genetics from the University of Wisconsin in 1943. She worked extensively in reproductive biology and population sciences, and became a lecturer in the School of Public Health at Harvard University (*American Men and Women of Science*). She was probably a biology technician at Los Alamos.

Frisch at least has left her name and a sketch of her life at Los Alamos. There are even more fragmentary records of other women, such as a nurse named Edith Tenney (Manley 1990). In general, like their counterparts at Hanford, female health workers at Los Alamos have left fewer traces than did those in physics and chemistry. Many other women must have worked in medicine at Los Alamos, but we have lost them.

A passage from Edith Quimby's obituary seems applicable to many female colleagues as well: "All too often, the creative achievements of scientific pioneers are overshadowed by further developments made by others or simply become anonymous components of accepted practice" (Rossi 1982). This was certainly the case during the Manhattan Project, when, because of security restrictions, much important research was not published for years, if at all, and scientists were not allowed to discuss their studies with the rest of the scientific community. Perhaps the records of accomplishment in this chapter will to some extent restore to female biomedical scientists and scientific workers of the Manhattan Project some of the recognition that they did not receive at the time.

# 7   The Technicians

THOUGH SKILLED scientific technicians rarely publish papers and few historians mention them, they play a critical role in most laboratories. They certainly did so in the labs of the Manhattan Project. Many of the technicians working on the project were women, filling jobs that during peacetime—especially in an economic depression—would have been taken by men.

General Leslie Groves, in overall command of the Manhattan Project, seems to have spent substantial time trying to recruit the skilled technicians he needed. He had to compete not only with other wartime research, such as the radar project, but with industries that needed skilled lab technicians for war-related work. It was no accident that the IBM machines at Los Alamos were serviced by an expert repairman: He had been drafted especially for the assignment.

Such blatant selective induction aside, the Army responded to shortages of technicians with several programs. The Special Engineer Detachment recruited soldiers with technical skills and assigned some of them to the Manhattan Project. The Army Specialized Training Program was designed to teach draftees technical skills. WACs with technical qualifications were likely to find themselves stationed at one of the Manhattan Project's major facilities.

Like their colleagues in the military, civilian technicians worked at all the Manhattan Project sites. Those whom we have succeeded in tracking down—like the scientists who worked on the project—tended to be young and idealistic, and they welcomed the opportunity to do scientific work.

The Met Lab hired Olga Giacchetti away from her job at the microphotographic laboratory at the University of Chicago in January 1945, assigning her to work in the Recovery and Solvent-Extraction Groups in the Chemistry Section. The part-time job meant more money and an opportunity to finish her undergraduate degree (Kathren et al. 1994). Giacchetti described her work during a symposium at Argonne in 1996:

My job as a laboratory technician was to count the prepared samples for alpha emissions, to make up necessary solutions, to always have available a clean supply of glassware, and to have plenty of lab supplies on hand.

No, I didn't have to make the coffee! We were in a restricted food area anyhow. I was instructed to heed the warning signs that were posted around the building: "Danger: Maximum working time 1 hour," or "No admission without changing shoes when going into the filter-air section," "Gloves not allowed in the counting room," and "Have you had your hands counted today?" They were very strict about these rules.

Six months later, Giacchetti switched to full-time work for the Solvent-Extraction Group and began chemical analysis of samples from the test columns. At first, like many Manhattan Project employees, she did not know what the project was about. "At the time I was hired I didn't know what research was being conducted, but I knew it was important," she said. "It wasn't till later that I learned we were working on research related to the ATOMIC BOMB!! I knew then, since the war was still going on, that the research being done was of the highest priority. You could just sense the seriousness of it!" (Fineman 1996)

Thanks to the Manhattan Project, Giacchetti continued her education at the University of Chicago. She received a B.S. in physiology in August 1946, and shortly thereafter she married Phillip Fineman, a former member of the SED (Kathren et al. 1994).

Female technicians joined the Manhattan Project for the same reasons as female scientists. In most cases their primary motivation was not the money, although for many of them a technician's job represented a step up. For example, records show that on March 5, 1945, "Betty Murray transferred from the position of typist to the position of technician for Lawroski. She will receive an increase of $0.75 per week" (Kathren et al. 1994, p. 632). By April, Murray was working half-time, and she was soon a full-time technician in the Solvent-Extraction Group.

Kathleen Gavin joined the Chemistry Section of the Met Lab as a clerk in September 1943. She took courses in chemistry at a community college in Chicago and became group supervisor of the Nonacademic Service Group, in charge of laboratory assistants, some technicians, and much secretarial work. Minnie Daniels, Aquilla Parsons, Elsie Mae Freeman, and Lillie May Porter were among the women whom Gavin supervised. In August 1944 she married Alan E. Florin, a chemist who worked next door. She left the Met Lab in May 1946, but later worked at Los Alamos (Kathren et al. 1994; Seaborg 1977).

Women's contributions to the war effort did not change society's view of them as sex objects. A contemporary collection of cartoons celebrating America's war industries, *The Home Front*, stated in the introduction that "America's war industries are a healthy, robust and productive lot. In fact they are the best in the world wearing a bandanna, pink sweater, size 36 brassiere and slacks." The Manhattan Project was typical of its times.

Helen Pellock joined the chemistry section at the Met Lab as a laboratory technician in April 1944, after attending Southern Illinois University. Glenn Seaborg mentions her twice in his journals:

> The fellows in Room 2 are still chuckling about a recent incident. Helen Pellock, a technician, is in the habit of taking off on her bike at lunch time for a swim. On a recent day she changed into her swim suit in the women's room, donned a pair of slacks, and then walked through the lab. As she passed Norm Davidson, who was working with some greasy thing, she got some of the grease on her suit just above the waist. Norm was quite distressed as it appeared to be his fault, so he tried to mollify her by cleaning off the grease. He took a wad of cotton, soaked in acetone, and wiped away the middle section of the rayon suit.

On September 23, 1944, Pellock married Sigfred Peterson, a Met Lab junior chemist. Peterson had interrupted his graduate studies at the University of Minnesota to go to the Met Lab, where he worked as a research chemist in Seaborg's section from 1944 to 1946. Seaborg recorded their marriage as "another laboratory romance" (Kathren et al. 1994, p. 526).

Female technicians came from all over the country, many ending up at secret locations far from their homes and families. Yvette Berry arrived at Hanford from Minnesota during the construction phase of the project, probably in 1944, and initially had nothing to do with the lab work. She had studied chemistry and physics in high school and had a year of college math, however, and Hanford recruiters offered her a job in the analytical laboratory. Berry missed the lakes and woods and felt little fondness for Hanford's heat and sandstorms. Even the recruiter admitted that the job would entail long hours and shift work. The pay was good, however, and the recruiter assured her that the work was not dangerous. Once she had accepted the job, she worked rearranging files while the FBI completed her security check. She finally joined the laboratory staff in February 1945.

Berry received two weeks of intensive training and then was allowed to analyze the "hot stuff." Like her coworkers, she would pipette a murky, greenish liquid from a vial under a hood and place it on a metal disk. She would clean up any extra material with cotton dental swabs held carefully in tongs. The disks were dried under a heat lamp, coated, and taken for radiation counting. Contamination was a constant problem for workers in this area, and they frequently had to spend hours scrubbing both the lab and themselves to remove invisible spots of radioactive contamination (Van Ardsol 1958).

Pearline (sometimes spelled Perline) Boykin, one of the few African American woman at the Met Lab, was born in Ofahoma, Mississippi, probably around 1911. She attended high school in Mississippi from 1932 to 1935, and beginning in 1940 spent three years at the Mississippi Nurses Training School. At the same time, she worked at the Mississippi Baptist Hospital, where she gained experience in using the X-ray machine.

Boykin left the South for Chicago and joined the Chemistry Section of the Met Lab in March 1944. She worked as a lab technician in a group working on process development, and in January 1945 became a technician in the Solvent-Extraction Group, which cooperated with the Industrial-Scale Semiworks, a half-scale plant built at Clinton. She worked in the Health Division at the Met Lab's successor, Argonne National Laboratory, after the war (Kathren et al. 1994).

Constance Simonsen grew up in Portland, Oregon. She spent two years at Reed College, and graduated from the University of Washington with a liberal-arts degree. In June 1943, she was working at a library in Portland when a friend, a chemist in Berkeley, called and asked whether she would like a job doing war work. Eager for a change of scene, Simonsen accepted without an interview. She arrived at Los Alamos courtesy of Army drivers, who loved to terrify newly arrived civilians. She was assigned to the Chemistry Division as a liaison between Chemistry and Procurement. In typical Los Alamos fashion, she shared an office with the head of the division, Joe Kennedy. Her job consisted of processing procurement orders for the division—and doing anything else that was needed, including working the switchboard from time to time.

The rugged life on the Hill bothered young, single staff members considerably less than it did the hard-pressed wives and mothers. Simonsen remembers working in little theater and organizing hiking trips. Since she was an early arrival at Los Alamos, she was initially housed in a dude ranch near the Tesuque pueblo. A lab bus—driven by an Army driver—ferried personnel to work in the morning and home at night. Simonsen met David Wood, an engineer and materials scientist, through a friend who was secretary to his boss. The Woods' wedding was the first of many held in Fuller Lodge. Otto Frisch supplied the music.

The liaison job eventually became boring, so in March 1944 Simonsen transferred to a job as a metallographic technician. Her group studied the effects of bringing pieces of uranium together at high speeds, in preparation for what would happen in the bomb itself. The uranium slugs were shot together, and the pieces were cut, polished, and photographed both with a conventional camera and through a microscope. Simonsen mounted and polished slides and developed film, always looking for deformations in the uranium structure. The hours were long and irregular, but the work was interesting and the democratic atmosphere lightened the load (Wood 1996).

Like the scientists, many of the technicians were very young. Project recruiters approached untrained women who showed aptitude for science and math, and offered them on-the-job training in advanced lab techniques as an inducement to work for the project. (The wives of Manhattan Project scientists, in contrast, frequently held at least an undergraduate degree in science or mathematics.) Several women recruited in this way went on to careers in science that they would never have dreamed of otherwise.

Jean Klein Dayton followed her husband, a physicist, to Los Alamos after completing two years of study at Cornell University in conservation and field biology. In order to have something to do and to qualify for maid service, she accepted the laboratory's offer of work. Help with the laundry aside, she was eager to get to know the local maids, some of whom were Native Americans who lived in nearby pueblos.

Dayton started in the Electronics Division, making Geiger counters and other equipment and installing an interoffice phone system. She transferred to weapons testing in order to get outdoors, and she later helped to design the detonation system for the hydrogen bomb. John von Neumann selected her for the job because he felt that a mathematician would take too long to figure out the system. A person working intuitively, he hoped, would be more efficient.

Although Dayton was aware that she was working on a weapon, she did not realize that it was an atomic bomb until the bombing of Hiroshima was reported in the newspaper. She left Los Alamos after the war and raised five children. She also worked at a wide variety of jobs, including lecturing on atomic energy at the American Museum of Natural History in New York and serving as a lab assistant at Sloan Kettering.

Dayton's career is an excellent example of how World War II opened doors to women. She was the first woman allowed to major in conservation at Cornell, one of the few women in the College of Agriculture, and the first to do such work for the New York State Conservation Department. Even at Los Alamos, trained technical workers were in short supply, and women entered fields where they would not have been allowed in peacetime (Dayton 1991).

Female technicians were concentrated in chemical-analysis labs at the Metallurgical Laboratory, Oak Ridge, and Hanford. At Los Alamos, many spent their time doing the endless calculations generated by the physicists of the Theory Division. Many female biomedical technicians worked at the Met Lab, particularly during the later years of the project. At Los Alamos and Oak Ridge, women worked as technicians in a wide variety of jobs. Wartime security prevented project participants from discussing their work with anyone, so women on the project, many of whom worked in otherwise all-male groups, did not always know what type of work their female friends were doing. For this reason and others, these technicians have proved difficult to locate, and those whose work is described below represent many who have escaped this history.

## Chemical Technicians at the Metallurgical Laboratory

Chemical research at the Met Lab required technicians as well as chemists. Many of the women who worked there stayed only a short time, and we know few details of their lives and work. Even Glenn Seaborg, head of the plutonium project, noted the unusual number of marriages between project staff members. Unfortunately for historians, most women changed their names when they married, making them difficult to track.

Women in the chemistry labs handled the new transuranic elements, which posed uncharted health risks. Because plutonium and uranium were expensive, much of the chemistry had to be done on microscopic samples. Separation processes required study of highly radioactive samples from uranium fuel rods that had been placed inside reactors. The processes that these women developed were recorded by the project leaders, but the women themselves received only brief credit in the laboratory annals.

Gayle E. Adams worked as a laboratory technician on three projects: investigation of the fission properties of U-233 and methods of separating it from large amounts of thorium; investigation of the chemical properties of elements 93 and 94; and the preparation of U-234 and measurement of its fission properties. Seaborg describes her as "an energetic technician" and "a jack-of-all-trades" (Kathren et al. 1994).

Helen Thomson was born in Baltimore in 1921 and attended Chicago Teachers' College for two years. She worked as a clerk at a small shop and as a technician at the Armour Company in Chicago. Thomson joined the Chemistry Section of the Met Lab and by November 1944 was handling Pu-239 as a technician. She eventually began working half-time, then full-time, as a technician in the Heavy Isotopes Group. She joined Argonne National Laboratory in 1946 and worked there until 1948 (Kathren et al. 1994).

Minnie M. Daniels worked as a clerk in the Women's Army Auxiliary Corps (WAACs) at Fort Clark, Texas, from December 1942 until August 1943. Born in Vienna, Georgia, in 1921, she attended Georgia State Normal College in Albany, Georgia. Daniels joined the Met Lab Chemistry Section in September 1943 and worked there until February 1944. At the time of her resignation she was one of Kathleen Gavin's assistants (Kathren et al. 1994).

We know these women's names and that most of them were young, and we know that they worked in the crowded labs near the University of Chicago. Lab records show the type of work they did and record marriages and accidents. But the records tell us nothing about their personalities, the day-to-day tension of performing exacting technical work for long hours, their concern about the toxic substances they were handling, or their anxiety about friends and relatives who were fighting overseas. Many of these women did not know the purpose of their hard work. They accepted the word of their supervisors that doing their job well would help to win the war.

The female technicians at the Met Lab must have watched the daily casualty lists with the same emotions as women elsewhere in the country, but the lab records don't note the deaths of husbands and fiancés. Nor do we know about divorces, despite the many references to project romances and marriages. We don't know what jokes kept spirits up in the labs or how the women handled the pressures of child care, rationing, and finding living quarters in a crowded city.

Ellen Watts worked as a technician in the Metal Production Group from March to October 1944. Beatrice Foreman did the same type of work from November 1943 to July 1944. Beginning in June 1944, Eleanor Lewis worked as a technician in the Instruments and Physical Measurements Group, where she handled the isotope niobium-93. She left the Met Lab in January 1945. Ruth A. Casler joined the Met Lab as a technician on December 1, 1943, and worked for several different groups during her time there (Kathren et al. 1994).

Shirley Nyden was born in Oak Park, Illinois, and attended Wilson Junior College for two years. She worked as a messenger for Butler Brothers and as a salesclerk for Sears and Roebuck. In June 1944, she joined the Fluoride Chemistry Group as a technician. She and three other staff members worked in Room 27, where in October 1944 dust and fume hoods gave off-scale "Pluto" (a portable alpha survey meter) radiation readings. Nyden left the lab in December 1944 to return to school (Kathren et al. 1994).

Selma Shupp was born in Palo, Iowa, in 1923. After two years at Coe College, she worked as a teacher in Palo. In June 1944 she joined the Met Lab in Chicago as a technician. She left in early September 1944, but reentered in late September as a technician in the U-233 Group. She left the project for good in January 1945. In February 1945, Marian Pinckard was hired as a technician in the U-233 Group, leaving at the beginning of July. Virginia Towle worked as a technician in the Instruments and Physical Measurements Group from March to July of 1945. Betty Mokstad worked in the Dry and Wet Chemistry Groups, also from March to July of 1945 (Kathren et al. 1994).

Sources mention a number of other female laboratory technicians in the Chemistry Division at the Met Lab: Adele Koskosky; Opaline Calhoun, who worked in the Control Analysis and Heavy Isotopes Groups in 1945–46; Winifred T. Koziolek; Virginia Meschke; and Emma Tolmach. The latter three women arrived in 1946 and worked in the Solvent-Extraction Group. Rachel Gilbreath came to that group from the Health Physics Group in April 1946 (Kathren et al. 1994), and Lillian

Guadagna joined the Met Lab in the same year, though it's not known what type of work she did.

Elaine Lewitz, who was born in Chicago in 1921 and attended the University of Chicago from 1939 to 1941, joined the Met Lab in February 1943 as a laboratory assistant and secretary to Elwin Covey in the Chemistry Section (Kathren et al. 1994; Seaborg 1977). In June 1943, she married Ted Novey, and the couple left Chicago for Oak Ridge in September 1943. They returned to Chicago in 1946, and Elaine worked at Argonne National Laboratory as a chemist, studying technetium (Freedman 1991). She worked in administration at Argonne from 1978 to 1989.

Kay Harvey was born in Larksville, Pennsylvania, in 1918. After high school she went to work as a lab technician for Du Pont, and she joined the Met Lab as a laboratory assistant in July 1943. At the end of August 1943, she transferred to the Clinton laboratories (Kathren et al. 1994).

Marilyn "Jodie" Jordan was a laboratory assistant at the Met Lab; she married Henry Hoekstra, a chemist, in August 1944. A number of other women, some with college training, some without, worked as lab assistants in the same group, among them Addie Saunders, Nellie Jennings, Eleese Bowman, Julia Elaine Freeman, Valda B. Lemke, and Evelyn J. Brown (Kathren et al. 1994; Seaborg 1979).

## FEMALE CHEMICAL TECHNICIANS AT HANFORD

The three main sites of the Manhattan Project handled very different types of work. Los Alamos dealt with bomb design and nuclear physics, and most of its workers were scientists and supporting technicians. Clinton Engineer Works housed not only experimental nuclear reactors and pilot separation plants but major production facilities for isotope separation. Its staff comprised construction crews and production teams as well as laboratory workers.

Hanford was initially primarily a construction site, and the town was in large measure a construction camp. Female employees were allowed to ride a bus into Pasco after work for shopping and other necessities, but the camp at Hanford was rough enough that a "women only" bus ran the return route on Saturday night.

Unlike Los Alamos, Hanford seems to have made no particular effort to recruit the wives of scientists stationed there. The vast majority of women worked as secretaries and clerks, moving the mountains of paper that the huge construction projects generated. Others staffed the mess halls and other facilities that served the army of construction workers. A company of WACs assisted with administrative chores. In contrast to Los Alamos, where the average age of the staff was thirty-one, 51 percent of the workers at Hanford were thirty-eight or older.

Married workers lived in rapidly constructed houses. Unmarried workers and the many people who had left their families behind to take temporary work on the project lived in single-sex barracks. The project maintained segregated barracks for black employees. The sixty-four women's barracks at Hanford could hold 4,480 residents; the 131 men's barracks held 24,319. Other workers were quartered in hutments and a trailer park. In return for these accommodations and a fifty- to fifty-nine-hour workweek, workers were promised the going wage and all the food they could eat. Nearly 600 employees quit each day (Powers; Van Ardsol 1988).

By January 1945, construction of Hanford's production reactors and separation plants was largely complete. It had been clear since June 1944 that the Metallurgical Laboratory and Clinton Engineer Works could not supply Hanford with enough trained people to carry out the necessary radiochemical analysis, so Du Pont managers decided to hire unskilled workers and train them. They needed workers who were not eligible for the draft—otherwise they were likely to be plucked away by the Army—which explains why seventy-five percent of the Hanford workers between eighteen and twenty-six years old were classified 4F. Women were another obvious choice.

In January 1945, Hanford's Laboratory Division began a recruiting effort to attract women to the chemical-analysis laboratories. Though the recruiting was primarily local, women came to Hanford from all over the country as word of jobs with high wages spread. Over the next six months, the division interviewed 650 women, hiring and training 109 of them. The women were assured that the work was safe and interesting, and the recruiters pointed out that doing shift work would mean days off during the week.

The women hired were between the ages of twenty-one and forty, with high-school educations. They had to be in good health, to be "alert and intelligent," and to have a "pleasing personality." Their training provided some instruction in safety and security along with the essentials of their work. Male employees generally received training at Chicago or Oak Ridge before transferring to Hanford (Crane and Switzer 1945), but most female employees were trained on-site (though at least one woman worked at Chicago before moving to Washington) (Van Ardsol 1958).

The analytical-chemistry laboratories, which were attached to the plutonium production facilities, measured the output of the separation "canyons"—vast tanks where the radioactive fuel rods were dissolved and the plutonium separated out. Employees who worked with or near this solution wore coveralls and canvas shoe covers when they reported to work, and carried film badges and gamma pencils (dosimeters) to monitor their exposure. On leaving the lab, they checked their hands with radiation counters. In the event of a spill, the lab supervisors oversaw careful washing procedures, including the use of a solution that took a thin layer of skin off the worker's hands.

Marge Nordman DeGooyer, a technical specialist in setting up and operating high-resolution mass spectrographs, took an unlikely route to a scientific career. Trained in bookkeeping and secretarial work, she tried to get a secretarial job with a taxi company in Hot Springs, South Dakota, after high school. Jobs were scarce and she wound up driving a cab instead. In 1944 she was happy to join her family in Washington, where her father was working on the construction of the Hanford facility. She applied to Du Pont for a job and was immediately employed as a secretary, but the interviewers told her that she had math ability and could get the equivalent of a college education if she took a job in the technical area. She agreed.

After a forty-minute bus ride to the site, Nordman was issued a film badge and two gamma pencils. She was curious about these devices, so she retreated to the ladies' room and took one of them apart. A spring immediately fell out, and she managed to put the device back together only with difficulty.

She was asked whether she would rather cook or sew. Nordman said she preferred to do neither, but would rather cook if she had to choose. That guaranteed her assignment to the analytical-chemistry lab, which performed work with wet chemicals. Women who answered "sewing," such as Yvette Berry, were sent to the counting room, where the alpha radiation from samples of plutonium was painstakingly monitored. Automatic timers had not yet been invented, so the job was slow and very boring.

In the analytical-chemistry lab, where she was one of only four women, Nordman learned to conduct the complex chemical analysis for determining what percentage of a given batch of material was plutonium. She learned chemistry quickly and became an expert at her job. Her work on the Manhattan Project led her to a career in analytical chemistry. She remained at Hanford until she was hired away by industry (DeGooyer 1995; Sorenson 1995).

Donna Robinson, an early recruit to the analytical lab, dropped out of business college in Salt Lake City when the war began. She worked for Remington Arms until she was transferred to Hanford to work in the Investigations Division. As the construction phase of the project wound down, her boss offered her a transfer to the Tennessee plant. She did not want to leave Washington, so he suggested that she apply for training in laboratory operations. Because she had studied accounting, she already had considerable knowledge of mathematics.

In October 1944, Robinson started laboratory training, which consisted of safety instruction and learning to pipette extremely radioactive materials in quantities of lambdas. (A lambda is one-millionth of a liter.) A pipette was attached to a glass syringe fitted with a piece of rubber tubing. The operator used a tiny amount of pressure on the tubing to control fluid uptake until it was just above the desired quantity. A cotton swab held with tweezers was then used to draw fluid out of the pipette until the precise amount needed for an accurate measurement remained. The contents of the pipette could then be diluted and the diluted material baked onto a metal disk for counting.

As soon as she completed training, Robinson was assigned to a facility preparing to monitor the separation of the first slugs from the production reactors. The facility conducted several weeks of "cold runs" before work on radioactive materials began in December 1944. Initial dilutions of highly radioactive solutions were made using a device called a Goldberg, after the artist Rube Goldberg. The analysis work went slowly at first, because workers frequently contaminated themselves with the solutions. Later the work became both faster and more routine, although security and safety precautions remained tight.

The lab operated round the clock. When a sample arrived, all the workers swung into action, but the samples came only at long intervals. On the swing and graveyard shifts, workers were allowed to pass the time playing cards, but on the day shifts they simply waited.

Donna Robinson felt that she shouldn't discuss her work with her family in Cokeville, Wyoming. Unfortunately, the FBI ran checks on workers, and Robinson's parents assumed that she had been involved in something criminal. She managed to convince them that the FBI was interested in her war work (Sorenson 1995).

## Female Technicians at Oak Ridge

The new, raw towns built for the Manhattan Project at Clinton Engineer Works stood in a sea of mud. Because Clinton began operations before housing construction could be completed, many workers lived away from the site and commuted to work on a fleet of buses. Like Los Alamos and Hanford, the Tennessee site had primitive housing, barracks for single workers, and a schedule dominated by the needs of the production plants and the labs.

Oak Ridge was home to the first reactor designed to produce plutonium and to the gaseous-diffusion plant for isotope enrichment. It housed the calutrons, giant mass spectrometers that performed the final enrichment of uranium samples. Lab workers called them "racetracks." Unlike Hanford, which was primarily a production facility, Oak Ridge conducted active research in collaboration with the Met Lab. More technicians were therefore employed at Oak Ridge, and they did a greater variety of work than at Hanford. Because they were not concentrated in a single laboratory, however, we've had more difficulty in learning what work they did.

Rosellen Bergman Fortenberg joined the laboratory at Oak Ridge in March 1944, the same year she received her B.A. in chemistry from Lawrence University. She was assigned to a group doing chemical analysis and got quick on-the-job training in the mechanics of laboratory work, which she had missed in college (Fortenberg 1991). Ada Kirkley, along with many of her classmates at Mississippi State College, was recruited in May 1944 by Tennessee Eastman Corporation, the company charged with establishing the uranium-separation plant at Oak Ridge. Perry had completed two years of a botany major. She became a technician at the Y-12 plant doing analysis of samples (Kirkley 1991). Bergman and Kirkley are typical of many technically trained women who worked on chemical analysis at Oak Ridge.

In 1942, Eleanor Eastin Pomerance joined the Radiation Laboratory at Berkeley, spending at least a year working on the calutron for enriching uranium isotopes. Pomerance's first husband had been a bandsman, and she had worked as a hatcheck girl and hostess at the ballroom where he played—a résumé that upset at least one of the stuffier physicists who interviewed her for war work.

In June 1945, Pomerance transferred to related work at the Y-12 plant in Oak Ridge. She had taken courses in engineering drawing at Berkeley High School in California, studied art for two years at Sacramento Junior College, then studied geology and anthropology at San Francisco Junior College. She spent two years as the only woman in the engineering-design section of Alameda Naval Air Station. After the war she joined Oak Ridge National Laboratory. Pomerance's most visible contribution was the design of the three-bladed fan in magenta and yellow that has become the standard warning label for radioactivity. She drew up the design with an Army noncom who was a chemical engineer (Pomerance 1991).

In 1944, Grace McCammon Estabrook joined Tennessee Eastman Corporation, the subsidiary of Eastman Kodak selected to operate the isotope-separation facility at Oak Ridge. Her job title was junior physicist. Estabrook held a bachelor's degree in science and mathematics from Maryville College and had just received her master's in accounting, statistics, and economics from the University of Tennessee. She was quickly moved from a job in the Accounting Department to the Statistics Division, where she was the statistician of a group charged with preparing top-secret reports on the operation of the Y-12 plant.

The electromagnetic-separation plants were so heavily classified that even output data were collected in code. Estabrook's job was to take coded data and prepare readable charts and graphs recording the operation of the plant. Only seven final reports were prepared each month. One went to the President of the United States; the others went to selected senior company and federal officials. After the war, Estabrook joined Oak Ridge National Laboratory and worked there as a physicist for the rest of her career (Estabrook 1991).

Barbara J. Smith had two years of science courses in college, first at Greenbrier College in Lewisberg, West Virginia, and then at the University of Tennessee. She dropped out of school because of severe migraine headaches and found work at the K-25 plant. There she operated a spectrometer, cleaned and repaired parts for it, and did calculations in a pinch. She enjoyed the work and the people she worked with, even though she had to commute by bus from Knoxville—at least a 25-mile ride—until housing became available at the site.

## FEMALE TECHNICIANS AT LOS ALAMOS

Experimental work at isolated Los Alamos covered a wide range. While some workers did careful analytical chemistry, others were field-testing explosives. In October 1944, 200 of the 670 civilian employees of Los Alamos were women, partly a result of General Groves's desire to make use of employees' spouses. Many of these women worked in the schools, and about two-thirds of them held jobs classified as administrative. Nevertheless, a substantial number of women worked as technicians in the laboratory divisions. At its peak in early 1945, for instance, the eighty-person Electronics Group included six women, who worked full-time making special cables for experiments (Hoddeson et al. 1993).

Many women received on-the-job training to supplement nonscience backgrounds. In April 1945, Kenneth Greisen, head of the group that installed the detonators on the implosion bomb for the Trinity test, needed extra workers to help test switches and other components of the detonation system. He requested the transfer of "3 more men of only modest or mediocre ability" and "4 more men or girls" as assistants (quoted in Hoddeson et al. 1993, p. 322). He seems not to have considered that women of "modest ability" might exist as well.

Many women, often wives of laboratory personnel, found employment in the Radioassay Group at Los Alamos, where they provided data for the control of the lab's precious supply of plutonium. In June 1945, Rebecca Bradford headed that group (Hoddeson et al. 1993). Bradford's home was in Pasadena, California, where her father was vice president of a bank. She obtained a bachelor's degree in physical education from the University of Southern California in 1941 and was knowledgeable in science at the undergraduate level. She was hired by the California Institute of Technology to do delicate work with quartz fibers on a project sponsored by the National Defense Research Fund: building oxygen-sensor systems for submarines. What she really wanted to do was travel to Mexico, but she didn't have the necessary $100.

After Pearl Harbor, Bradford found herself locked into her job, which became routine as the project entered production. When a former coworker requested her release so she could continue work with quartz fibers at Los

Alamos, she was eager to go. She started there in 1943. Her sister's boyfriend learned about the secret project and wrote to say that he knew she was working on an atomic bomb. The FBI evidently intercepted the letter and showed up to question Bradford about the security leak.

At Los Alamos, Bradford worked on quartz-fiber balances for determining the mass of microgram quantities of plutonium. The mechanism never worked properly, because slight electrostatic charges slammed the delicate quartz system against the side of the container and broke it. The group made only one successful measurement. Bradford then trained to do radioassay work on plutonium samples and to calibrate samples of uranium from weapons tests. She remained at Los Alamos after the war, eventually leaving to start a family (Divan 1991).

A number of female technicians worked for the Medical Group at Los Alamos. Jeanne Carritt held a bachelor's degree in biology from Oberlin College (1936) and a master's from Smith (1938). In June 1944, she accompanied her husband to Los Alamos, where he was assigned to chemistry work. Carritt had no job offer when she arrived, but planned to seek employment. She took an "on-the-job training" position in the Metallurgy Division.

Carritt had some experience in photography and put it to use in the Metallurgy Division, but as soon as the Medical Division was established she asked for a transfer. She was soon promoted to a staff position in the Medical Division, but her salary remained fixed—an injustice about which she complained for the rest of her tour at Los Alamos. In the Medical Division she worked with Dr. Elizabeth Maxwell on radiation-testing programs and on research into the effects of radiation (Carritt 1992).

Marilyn McChesney, described by coworkers as a real Marilyn Monroe type, worked in the Medical Division collecting blood and urine samples. Since all personnel handling radioactive materials were supposed to undergo tests once a week, the work absorbed considerable time and energy.

Other women filled less conventional roles at Los Alamos. Frances Dunne, who describes herself as "more or less an orphan" from the age of four, attended various colleges, including Swarthmore, but never got a degree. Instead she earned a license as an aircraft mechanic and was working as a senior mechanic at Kirtland Field in Albuquerque when the war began. Her mechanical ability—and her small hands—caught the attention of George Kistiakowsky, who was then in charge of explosives work at Los Alamos. He hired her for work with the Explosives Group.

The group, which was testing scale models of various bomb assemblies, needed a skilled person to reach into a tiny opening to set the triggering mechanism, which was surrounded by live explosives. The work demanded considerable mechanical ability and cool nerves. Dunne was the only woman to work at the explosives sites: alpha, beta, and Two Mile Mesa. Her group consisted of herself and thirteen tech sergeants from the Special Engineer Detachment, whom she still calls "the boys."

Dunne did not realize until 1945 that she had been working on an atomic bomb, even though her group did the explosives testing for both Little Boy and Fat Man—the Hiroshima and Nagasaki bombs, respectively. As explosives supervisor, she would assemble the trigger, then serve as the technician in charge of the test. She did the countdown and pressed the button that detonated the charges. The mockups of the bombs were tested for air, land, and sea delivery by running them through a large tank of water and shooting them off a wooden frame.

Dunne's group was the final assembly team for both bombs used in the war, but she was not allowed to go to the South Pacific with the rest of the group for the final work. As the time for the actual use of the bomb approached, a B-29 was flown to Kirtland to test loading procedures. The only vehicle available to transport the mockup from Los Alamos to Albuquerque was a new C2 wrecker. To avoid attention, the truck made the trip over back roads at night. Dunne, who knew the roads, made the trip with the mockup.

Dunne, who as a civilian could argue with Army officers without fear of discipline, was also called upon by her military group when it was necessary to deal with Los Alamos security personnel. One night, for example, she was awakened and asked to persuade the gate guards to admit a general who was wearing civilian clothes and whose pass had expired.

After the war, Dunne left explosives work in favor of a career with the FBI (Dunne 1991).

## The WACs

In the face of wartime personnel shortages, the Manhattan Project called on enlisted men and women to fill essential positions for which civilians were unavailable. Members of the Women's Army Corps were assigned to all three major project sites between 1943 and 1945. Although many worked as drivers and clerks, some were scientists and technicians (V. C. Jones 1985). Norma Gross, for instance, whose work has been described in the chapter on chemists, was a WAC.

In May 1945, sixty-seven WACs were stationed at Los Alamos, about 40 percent of whom worked in the scientific areas—a higher percentage than the figure for all Los Alamos staff. Probably the most accomplished scientist among the WACs stationed at Los Alamos was Mary Lucy Miller, a physical and polymer chemist. Miller received her bachelor's degree from Syracuse University in 1927 and her master's and Ph.D. from Columbia University in 1929 and 1934, respectively. She taught at St. Mary's Junior College in North Carolina in 1931–32, and worked as a chemist at Rockefeller University from 1934 until 1938. She was a research associate at the school of medicine at Washington University in St. Louis when the war broke out (*American Men and Women of Science*).

Miller enlisted in the Army in 1943 and was stationed at Los Alamos. Project administrators recognized her talent and qualifications and placed her in charge of one of the chemistry labs. She was only a private, and though she was offered a military rank suitable to her new responsibility, she refused all promotion as a matter of principle, though she seems to have welcomed a leadership role in the laboratory. (She remained in the WACs until 1945 (*American Men and Women of Science*). After the war she became a research fellow at the American Cyanamid Company, continuing to work there as a chemist until her retirement in 1969.

Another WAC chemist, Elizabeth Wilson, learned radiochemistry at Los Alamos and later pursued a career at Brookhaven National Laboratory (Spindel 1993). Evelyn S. Walker joined the Army in March 1943. She was assigned to the Manhattan Project, where she eventually did work with metal oxides and plastics (E. S. Walker 1991).

Myrtle Bachelder held an undergraduate degree in chemistry from Middlebury College and an Ed.M. from Boston University. She worked as a teacher in South Hadley Falls, Massachusetts, and joined the Army in 1942. After training in San Francisco, she attended Officer's Candidate School in Des Moines and received a commission as a second lieutenant in the WACs. She was given secret orders to take a company of fifteen to twenty women to Fort Sill, Oklahoma, and then to Santa Fe.

The WACs under Bachelder's command were fitted into various slots on the project, apparently at random. Bachelder herself was assigned to a section that used analytical chemistry and emission spectroscopy to determine the purity of materials used in the preparation of uranium and plutonium. She learned spectroscopy on the job. She was invited to view the Trinity nuclear test, but was on leave when it took place. After the war, Norman Nachtrieb, the chemist Bachelder had worked for at Los Alamos, brought her to the University of Chicago, where she continued to work in analytical chemistry (Bachelder 1991).

Lyda Speck had finished all the science and math requirements for a bachelor's degree when finances forced her to leave school. She worked for the U.S. Postal Service for four years, then took a leave of absence from her civil-service job to enlist as a private in the Women's Army Corps. After basic training in Orlando, she was sent to Los Alamos in November 1943. Working in the Experimental Physics Group, she became the only woman to work with the Van de Graaff accelerator that had been transported from the University of Wisconsin. Her job was to develop the photographic emulsions and make the thousands of measurements of tracks needed to determine neutron energy.

In September 1945, Speck received a letter from the laboratory director, J. Robert Oppenheimer, commending her on her contributions to "some of the fundamental research problems in this laboratory." It said in part: "For the past twenty months you have worked as an assistant in our research laboratory making microscopic measurements which called for a great deal of judgment on your part. This work was exceedingly tedious and involved a good deal of nervous strain. Nevertheless you have performed your duties in a cheerful and diligent manner, and it must be clear to you that you have made a real contribution to the success of the project" (Speck 1991).

Speck was promoted to sergeant while at Los Alamos, and her work formed the basis of two papers published as abstracts in *Physical Review*.

Miriam White Campbell studied architecture at the University of Illinois before joining the Army in 1943. She had had a part-time job with the U.S. Geological Survey, working on drawings for a machine to manufacture charcoal briquettes. In the military she was offered a chance to go to Officer's Candidate School, but her superiors decided that it was more important that she go to Los Alamos. There she was assigned work as a draftsman. Her colleagues were all civilians, and military rank meant very little. Everyone used first names, and frequently didn't even know one another's last names. Fortunately, Campbell's boss, an escaped Russian, felt that good work should be rewarded and tried to get her a new stripe every six months.

One of Campbell's jobs was to draw the detailed plans for the assembly of Little Boy, the uranium-gun bomb dropped on Hiroshima. She worked on the project in the office she shared with five other draftsmen until an Army captain stopped by to check on her progress. He was horrified to find such secret work being done in relatively open surroundings, and he pulled her out of the group office and assigned her a desk in an office kept for John von Neumann until she had completed the project (Campbell 1991).

Harryette Hunter Emmerson received a degree in home economics from Oklahoma College for Women in 1930 and quickly qualified for Officer's Candidate School after enlisting as a private in 1943. She was sent to mess school and was eventually assigned to Los Alamos, where she served as company commander and a dietitian at the hospital. She was invited to attend the Trinity test but refused, as her sergeants had not been invited. She did see the site the day before the test and returned to it the day after. Allowed to walk into the crater carrying a Geiger counter, she was impressed by the fact that the blast had turned the desert sand into glass of different colors, depending upon its impurities (Emmerson 1991).

The female technicians we have been able to locate often downplayed the difficulty and danger of the exacting jobs they performed. Lab stories tend to stress amusing accidents and visits by dignitaries rather than the day-to-day routine of experimental work. Still, even the brief descriptions we've been able to provide show how critical the work of these technicians was to the construction of the atomic bomb.

Each technician performed only one small step in the development of a nuclear weapon, and their work was reported by their supervisors, most of whom were male scientists. The stories of these technicians, male and female alike, have been largely left out of the history books. They deserve better, because the broad theoretical schemes of the project's leaders could never have developed into usable weapons without their efforts.

# 8  Other Women of the Manhattan Project

ONE WOMAN who worked on the Manhattan Project in New Mexico drove a five-ton truck at the Trinity construction site. History has not recorded her name, but she is remembered for knocking down the only man who knew the layout of the 500 miles of electrical lines installed for the first test of the atomic bomb (Bainbridge 1974).

Not all of the women who contributed to the Manhattan Project worked on its scientific aspects. There were truck drivers, switchboard operators, and, above all, secretaries and clerks. The Manhattan Project moved tons of supplies and fought titanic battles with military procurement and security. In this regard, the project typifies the U.S. war effort. One World War II recruiting advertisement read, "It takes twenty-five girls behind typewriters to put one man behind the trigger in this war" (quoted in Brinkley 1988, p. 108). That was probably a considerable exaggeration, but the demand for clerks was real. The salary, $1,440 a year, looked very good to women long out of work in the Depression—though it was, of course, considerably less than what men were offered for comparable jobs.

Many of the women who were called secretaries on the project carried major administrative responsibilities and held top-level security clearances. "Initially I did plan on having an executive officer," wrote General Leslie Groves, head of the entire Manhattan Project, "and I selected first one and then a second highly competent man for that position. Yet, before either could even begin to take over his duties, I had to reassign him to fill a pressing need in the field, one at Hanford and the other at Los Alamos. I soon realized that as long as we were under such pressure I would always find it necessary to assign to the field anyone whom I might consider acceptable as an executive officer in my headquarters. Consequently, I abandoned all further attempts and relied instead upon my chief secretary, who became my administrative assistant. With her exceptional talents, and her capacity for and willingness to work, Mrs. O'Leary more than fulfilled my highest expectations" (Groves 1962, pp. 28–29).

Jean O'Leary was only thirty when she moved to Washington with her small daughter in 1942. She had worked as a secretary at *Time* magazine, and was hired quickly by the War Department. After a few months she was assigned to Groves, who was still in charge of the construction of the Pentagon. The general was rough on secretaries, and he expected them to exhibit the same devotion to their jobs that he demanded of himself. He showed a complete lack of consideration for his subordinates' feelings and almost never dispensed praise. Nowhere in his book does he even mention O'Leary's first name. O'Leary, however, was able to cope with such a boss. She was struggling to recover from the trauma of her husband's recent death, and was willing to work long hours at a demanding job. She also possessed a sense of humor and an appreciation of the general's formidable ability (Groueff 1967). She needed both.

Arthur Holly Compton, director of the Metallurgical Laboratory and an administrative and scientific leader of the Manhattan Project, called his secretary, Kay Tracy, "in some ways the most important" of his three chief assistants (the other two being the male physicists in charge of administration and recruiting). "It was a matter of rare good fortune," he wrote, "that from the time the atomic program began to occupy my attention I had as my secretary a person perfectly fitted for meeting with discretion and good humor the many trying situations that arose" (Compton 1956, p. 86). Tracy, who had relatives fighting in the Pacific, supported both the development and the use of the atomic bomb (Nichols 1987).

Two other important secretaries worked for Colonel Kenneth Nichols, District Engineer for the Manhattan District. Virginia Olsson in Tennessee and Anne Phillips in New York managed offices, communications, travel schedules, and classified documents. They were valued legacies from Nichols's predecessor (Nichols 1987).

Many of the scientists involved in the Manhattan Project acknowledge the support of their wives, who obviously played significant roles in their husbands' work. For example, after mentioning his wife's efforts to find housing for Met Lab employees, Arthur Compton wrote, "Over the years my wife and I had worked together on various scientific enterprises, and now, though not a part of the technical councils, she was in spirit a part of the atomic project" (Compton 1956, p. 115).

The wives, of course, faced their own challenges every day, trying to maintain some semblance of normal family life in the turbulent atmosphere of the project sites. They missed their husbands, who were working long hours in the labs and shops, and those who were mothers worried constantly about the sort of schooling their children would receive. At all three major sites, administrators report spending incredible amounts of time on such issues as fresh milk supplies. Only the women's knowledge of the urgency of the wartime effort kept complaints at a manageable level.

At Los Alamos, which offered the young scientists and their wives little outside entertainment, the birthrate exploded long before the first bomb did. The first year saw eighty births, most of them first children, and the site averaged ten births a month throughout the war (Barnett 1988). In June 1944, one-fifth of the married women at Los Alamos were pregnant, and children, one-third of them under the age of two, accounted for a sixth of the population (Manley 1990).

General Groves was evidently not pleased by this unexpected strain on the project's limited medical facilities. An unknown poet recorded his reaction:

> The General's in a stew
> He trusted you and you
> He thought you'd be scientific
> Instead you're just prolific
> And what is he to do? (quoted in Marshak 1988, p. 16)

At Hanford, families were outnumbered by single construction workers and clerical staff. Despite the fact that the women's dormitories were surrounded by a high chain-link fence, the military found it difficult to uphold the desired standards of conduct. Los Alamos had no chain-link fences, so the post commander stationed MPs in the dorms. This incensed the residents, most of whom had graduated from college, if not graduate school, and felt no need for military chaperons. In a heated debate before the Town Council, the Army representative mentioned a rise in the syphilis rate, shocking the woman representing the women's dorms into tears. For some reason, the machinists, who usually vigorously resented any infringement of their freedom, were lower-key. As one of them put it, "The MP's a nice guy. If he looks lonesome, we give him a beer" (quoted in Smith 1988, p. 78).

Relations between the women who were concentrating on their families and the women who were working on the project tended to be edgy. This was the case even at Los Alamos, where many wives were pressed into at least part-time work. For example, Eleanor Jette, who refused on principle to be recruited for any kind of work at the lab, wrote of Joan Hinton:

> She played either fiddle or viola; which ever it was, she played well and joined the music lovers. She was a congenital twitcher, human fly type. The night she first played at our apartment, she wore a skirt and did everything but climb the walls when she wasn't fiddling. When the musicians left, I asked,
> "Does that girl ever sit still? Does she behave that way at work?"
> "When you're talking with her, she's just as likely to end up on the back of a chair with her feet on the seat as not," Eric [Jette's husband] admitted. "She's a smart cookie, but she had a progressive education."
> "Too progressive, if you ask me. I hope she wears slacks to work."
> "Not always; it gets pretty embarrassing." (Jette 1977, p. 62)

The feelings from the other direction could be similar. Charlotte Serber describes the resentment of working women at Los Alamos when nonworking wives began to receive more and more of the scarce household help: "Slowly, the working women began to realize that this service was no longer run primarily for their benefit, but for the community as a whole, and that in a way they were losing ground. We had no time to haunt the Housing Office to ask special favors" (C. Serber 1988, pp. 68–69). The Housing Office tried to calm things by instituting a priority system, with illness, pregnancy, and full-time employment at the top. This worked until an extreme labor shortage reduced working women's allotment of household help to a maximum of two half-days a week. Some women threatened to strike unless more help became available, though Serber says that they were bluffing. "We had been on the Project from the early days, we all felt a personal responsibility for its success, and it would have taken a great deal more than this minor crisis to make us quit" (C. Serber 1988, p. 69).

In this chapter, even more than others, we treat only a very few of the many women whose work helped to keep the war effort going. We've briefly mentioned the secretaries and drivers, and there were many others, such as the schoolteachers, who battled primitive conditions to educate the children of the community, and the wives and mothers not directly affiliated with the project, some of whom lived for months in tents while their husbands toiled long days on the manufacture of the bomb. They kept the daily life of the project communities going, and their role in the production of the bomb was real.

This chapter will tell the stories of Dorothy McKibbin, Charlotte Serber, Priscilla Greene Duffield, and Buena Maris, who played critical and well documented but unconventional roles in the Manhattan Project. We'll then discuss the roles of the WACs who were not scientists or technicians, and look at several other female workers, each typical of her colleagues.

## Dorothy McKibbin

Staff members arriving at Los Alamos were generally told to travel to Santa Fe and to report to office number three at 109 East Palace Avenue. The confusing streets of the old Spanish city led incoming scientists and technicians, and their weary families, to a small office marked with a simple red-on-blue sign reading "United States Engineer Office No. 3."

There they met Dorothy McKibbin, who was responsible for providing them with passes and directions up the twisting roads to the mesa on which the laboratory was located. McKibbin served as a general font of information and help for the sometimes lonely and confused young staff of the lab. She enforced security when project staff members came to Santa Fe, and disciplined everyone from GIs to Nobel laureates (McKibbin 1988).

McKibbin, a graduate of Smith College, first arrived in Santa Fe in 1926 as a tuberculosis patient. She had broken off her engagement when she learned of her illness, but her quick recovery led to marriage and life in St. Paul. Her husband died in 1931, and the following year McKibbin returned to Santa Fe with her son, Kevin. She earned fifty cents an hour working at the Spanish and Indian Trading Company. When the company shut down in 1943, she accepted a position as a War Department secretary with unspecified duties after meeting Robert Oppenheimer (McKibbin 1980). She knew only that her job concerned a very important war project (McNulty 1946).

In March 1943, 109 East Palace opened Site Y of the Manhattan Project. Santa Fe is still laid out around a central square, the Plaza. Local stores and offices occupy adobe buildings. The four offices of the Manhattan Project lay down an alley behind the stores fronting the Plaza. This made things easier for the security men, but it caused considerable difficulty for newcomers.

Oppenheimer and his secretary, Priscilla Greene, occupied one of the offices. A second belonged to D. P. Mitchell, who ran procurement for the project. The third went to J.A.D. Muncy, a fiscal representative from the University of California, who maintained the third-largest bank account in Santa Fe in order to pay hourly employees without recourse to the War Department. Security dictated that the name "Los Alamos," the word "physicist," and the title "Dr." never be used. The address for the entire laboratory was a post-office box—a box that certainly received some unusual equipment. As the project continued, numerous children were born inside Box 1663 (McNulty 1946).

In spite of the emphasis on security, the citizens of Santa Fe were well aware that something was up. One weary family looking for the Manhattan Project office stumbled into a bakery instead. As they stood staring in helpless confusion, expecting anything from codes in loaves of bread to armed escorts, a bored voice announced, "You must be looking for the office down the way; people are going in and out of there all the time" (Fisher 1985, p. 29). Santa Fe merchants automatically delivered large purchases by out-of-town buyers to 109 East Palace without even asking for an address (Fermi 1954, p. 234).

In the first hectic months, arriving scientists could not live at the laboratory site because the housing hadn't been completed. The project rented four dude ranches, and personnel were bused to work each day. Working with Rose Bethe, the wife of a theoretical physicist, who had undertaken the management of housing at the site, McKibbin ordered box lunches for the workers. She was careful to use different restaurants around the city so that no one would grow curious about the sudden increase in picnics. By May 1943, construction was far enough along for the rest of the staff to move to the site, and McKibbin went out of the box-lunch business (McKibbin 1980).

As the project grew, McKibbin dealt with a steady stream of problems: pacifying WACs who had thought they were being sent overseas and couldn't understand what they were doing in the middle of New Mexico; arranging for the repair of worn-out telephone lines (a constant irritation); keeping the project's buses from blocking local traffic; overseeing the accommodations for employees who had traveled to Santa Fe to apply for jobs and had to wait in local hotels for two long weeks while security checks were completed (McKibbin 1988).

In the summer of 1944, McKibbin was preparing Frijoles Lodge in Bandelier National Park to serve as temporary housing. The lodge manager was extremely unhappy about this plan, because the construction workers who had lived there in the summer of 1943 had left the place a mess. McKibbin cleaned up the lodge and hired a chef—only to learn that the expected personnel wouldn't be coming after all. Ever adaptable, McKibbin turned the lodge into a recreation center for the project staff. She was glad to return to Santa Fe at the end of the summer (McKibbin 1980).

Los Alamos even expected McKibbin to deal with an attack by Japanese balloons. A few balloons had indeed been dropped over the Southwest, and Los Alamos administrators panicked. They called McKibbin and asked her to drive to the top of a local hill and scan the sky for danger. She saw nothing but clouds, though her imagination turned those into a variety of balloons and parachutes. Other staff members on the mesa, however, spotted a bright object and spent the day staring anxiously at the sky. The Army sent up search planes. Fortunately, Los Alamos's personnel director was an astronomer, and he explained that the threatening object was the planet Venus. The staff went nervously back to work, not entirely reassured until the object appeared at the same place the next day (McKibbin 1988).

Although McKibbin was not officially a member of the site's technical staff, all aspects of the project crossed her desk. She watched the Trinity explosion from her car on Sandia Peak in Albuquerque (Szasz 1984).

Project staff members were not allowed to travel far from the site and could have no social contacts within 200 miles of Los Alamos, even if they had grown up in Santa Fe. Those who were occasionally able to meet relatives claimed that they could always spot the security personnel because they were so much better-dressed than the natives.

McKibbin allowed at least thirteen couples to use her house for their weddings. She could not tell the presiding judge or minister the full names of the couple, so the ceremonies were performed using only first names (McKibbin 1980). On at least one occasion, Robert Oppenheimer and Captain William S. Parsons, head of the Ordnance Division, attended the ceremony. McKibbin's neighbors were not pleased to be stopped on their way home by armed guards (McNulty 1946).

Almost all stories of life at Los Alamos recount episodes of McKibbin's resourcefulness. "All women brought their difficulties and their checks to Dorothy," wrote Laura Fermi. "She indorsed the latter so they could be cashed at the bank, and smoothed out the first: Yes, she knew of a boys' camp; Yes, she could arrange for a ride to the mesa later in the evening; Yes, she would try to get reservations at a good hotel in

Albuquerque; Yes, she could give them the key to the ladies' room. Women always came out of Dorothy's office with greater cheer than when going in" (Fermi 1954, pp. 234–235).

In its introduction to an interview, the official Los Alamos magazine described McKibbin as follows: "She issued passes, cared for children, rescued the lost and disillusioned, made housing and transportation arrangements, forwarded freight, and kept an ear to the ground for espionage agents. Perhaps more than anyone else, she earned the title Los Alamos' 'first lady'" (McKibbin 1980).

## Priscilla Greene Duffield

Priscilla Greene was working as a secretary at the Berkeley Radiation Laboratory when the director, Ernest Lawrence, lent her to Robert Oppenheimer to help organize the fast-neutron project, which grew into Site Y. She typed Oppenheimer's correspondence and became so infected by his enthusiasm for New Mexico that she quit her job at Berkeley and moved to Los Alamos. Greene arrived in Santa Fe in March 1943, along with the Oppenheimers' two-year-old son and his nurse (Smith and Weiner 1980).

Greene found Site Y "a pretty appalling place. It was windy, dusty, cold, snowy... and nothing was finished" (quoted in Hoddeson et al. 1993, p. 62). She moved into the confusion of the temporary project offices at 109 East Palace. She purchased her own typewriter in Santa Fe and spent the rest of the war trying to get reimbursed (Hoddeson et al. 1993). The Santa Fe office was thirty-five miles from the Los Alamos site, but the executive offices had to remain in town until a working telephone connection to the lab could be established. The original iron wire to the mesa lay along the ground and connected to a crank telephone. Charlotte Serber, who had been hired as scientific librarian and was helping out at the office, wrote: "Once a call came from the Hill asking us to send up eight extra lunch boxes. The request as we heard it above the noise, but lucidly could not fill, was for eight extra-large trucks" (C. Serber 1988, p. 59).

In those early stages of the project, each pass had to be typed in triplicate and signed by Oppenheimer personally. The mail had to be collected and shipped to different forwarding addresses in order to conceal the project's location. Greene and her colleagues, a collection of working wives drafted into the project, learned how to order and run mimeograph machines and telephone switchboards (C. Serber 1988).

Once at Los Alamos, Greene managed the director's office. She welcomed visitors, answered the telephone, and took notes on phone calls unless General Groves ordered her off the line. She typed correspondence, managed the files, and dealt with travel and security arrangements in a dusty military encampment with temporary buildings and a staff of temperamental physicists and engineers. She apparently managed to stay reasonably calm; at least, there are no reports of her losing her temper.

Greene's wedding to Robert Duffield, a chemist, was one of those that took place at Dorothy McKibbin's house. Oppenheimer could not attend the wedding because of a sudden meeting in Wyoming, but he did ask her to drive him to Santa Fe. He was worried about marriages in the pressure of the wartime environment and wanted to give her some fatherly advice (Smith and Weiner 1980).

## Charlotte Serber

In April 1942, Charlotte Serber and her husband moved to Berkeley from Illinois at Robert Oppenheimer's request. Robert Serber, a theoretical physicist, had been one of Oppenheimer's students at Berkeley before leaving for the University of Illinois. Back in California, the couple lived in a small apartment over Oppenheimer's garage. While her husband began work on the nuclear project with Oppenheimer, Serber took a job in a shipyard, where she was called a statistician because she could do simple mathematics (R. Serber 1996).

In March 1943, the Serbers moved to Santa Fe with the first wave of Manhattan Project scientists. Charlotte Serber had been hired as scientific librarian for the project. As such, she was responsible for the top-secret technical materials exchanged among the project's sites. During the first months, however, the library had no books, so Serber assisted Priscilla Greene in the director's office. She recalls nearly being killed during her first two weeks at the site—by a telephone. The new phone lines had no lightning arresters, and as she was reaching for the phone, a bolt jumped from it to a light, smashing the light completely.

One of Serber's responsibilities was handling the mail. All mail, incoming and outgoing, went through a number of post-office boxes in different locations around the country to conceal the fact that a large number of scientists were assembled at Los Alamos. A single secretary at each location forwarded the mail to the Santa Fe post-office box. One of the chemists' wives had been drafted to make two long trips down the winding, cliff-edge, dirt road to Santa Fe each day to collect the mail, accompanied by an armed guard. The mailbag was locked to her wrist. Only Serber had a key, so the young woman would report to Serber—usually at lunchtime or dinnertime—to be released. If Serber did not distribute personal mail immediately, she later wrote, she risked being ostracized (C. Serber 1988).

Security complicated a lot more things than mail delivery. A series of nightly inspections insured that classified materials were locked up. Violators had to pay a high fine or work as inspectors themselves. The system worked well once the laboratory had more than one safe; at first, only the director's office had one. Only one scientist escaped punishment for leaving a classified report on his desk. He argued that since the report was completely wrong, giving it to the enemy would be a service to the war effort.

When the first load of top-secret documents was delivered to the library, the vault was still under construction. The staff substituted an ancient safe with a special security feature: It opened only if Serber gave it a hard kick at a critical point while dialing the combination (C. Serber 1988).

Serber boasts of her brief service in counterespionage. Oppenheimer decided that rumors about the laboratory were getting out of hand: One story, based on the presence of medical equipment and a number of "doctors," was that the lab was a home for pregnant WACs; another persistent story held that it had something to do with submarines. Though these stories were comfortably far from the truth, Oppenheimer worried that the arrival of a cyclotron and other such clues would lead somebody closer.

Oppenheimer decided that Charlotte Serber and John Manley, an experimental physicist who was Oppenheimer's deputy, would make frequent trips into Santa Fe, where, pretending to be slightly drunk, they would plant the rumor that the Hill was making an electric rocket—a story that, had it reached the enemy, would have been dismissed as utterly loony. Serber asked whether she could explain the plan to her husband so he wouldn't be jealous, and Oppenheimer immediately added Robert Serber and Priscilla Greene to the team. The group diligently traveled into Santa Fe and spent a frustrating evening in the bar at the LaFonda Hotel, where nobody seemed even remotely interested in Site Y. The exhausted group returned home and firmly announced their retirement as secret agents (C. Serber 1988).

In his thank-you letter to Serber, written on November 2, 1945, Oppenheimer wrote in part:

> I think no single hour of delay has been attributed by any man in the laboratory to a malfunctioning, either in the Library or in the classified files. To this must be added the fact of the surprising success in controlling and accounting for the mass of classified information, where a single serious slip might not only have caused us the profoundest embarrassment but might have jeopardized the successful completion of our job....
>
> Certainly it would be right that you take pride in a job well done—an integral part, and a most important one, in this strange, but rather heroic, undertaking. (Smith and Weiner 1980, pp. 313–314)

## BUENA MARIS

Buena Maris filled a role at Hanford as unusual as Dorothy McKibbin's at Los Alamos. General Groves recruited her to Hanford from her position as Dean of Women at Oregon State College in September 1943. According to Groves, her job was "to deal with the problems that are bound to arise when a large population of women are housed as a group in an isolated area under rugged conditions, with few of the amenities of normal life" (Groves 1962, p. 90). Called Supervisor of Women's Activities and paid $4,000 a year, Maris was the only woman who attended the weekly senior staff meetings at Hanford (Williams 1993). Like McKibbin, Maris held a job that defied simple description but was critical to the success of the Manhattan Project. Her mission was to reduce the turnover rate of female employees, which ran as high as 20 percent a year.

The women's barracks were located in the Hanford construction camp, eighteen miles from the nearest town. Young women lured by promises of important war work in the glamorous Pacific Northwest found life in Hanford's desert environment disappointing, to say the least. Barracks life in a construction camp offered few amenities, although all agree that the food was very good and plentiful, an important consideration for Depression-era workers. Typists earned a munificent $26 a week. Cooks got $40, lab assistants $41, and experienced nurses $50.

The women represented many age groups, educational backgrounds, and races. Georgia Mason Coatie left her seven children with their father for the duration of the war. Drawn by the promise of well-paid work,

Coatie, an African American, traded her Midwestern home for the Hanford barracks. Geneva Owen Hammer, seventeen years old, was a coffee girl at a mess hall. She came to Hanford primarily to be near her parents, and she shared a barracks room with her mother (Williams 1993).

Groves and the military commander at Hanford, Colonel Franklin Matthias, supported Maris in her work. She hired "house mothers" for each women's barracks, organized a library, and started a Red Cross chapter. After riding back from Richland on a bus full of inebriated construction workers, she insisted that a late bus, for women only, be added to the schedule. She strongly supported the work of the recreation office in starting sporting events, dances, and even beauty contests, and she founded a Girl Scout troop for the camp (Gerber 1993).

Maris even conquered the gravel walks around the barracks. The gravel destroyed shoes, which were irreplaceable during wartime rationing. In the summer of 1944, Du Pont's chief engineer at the camp, G. M. Read, asked Maris what her greatest need was. She asked for sidewalks, or at least asphalt, to settle the dust and save the women's shoes. To her surprise, workers put down asphalt the very next day. On another occasion, workmen showed up to install a new roof on a vacant church and to perform other repairs that Maris had requested while Maris was holding a meeting in the building. The construction didn't even slow her down: Both the scout troop and the Red Cross chapter were organized at that meeting (Groves 1962).

Nothing stopped Maris. She persuaded a well-known clothing store to open a branch in nearby Pasco. She obtained a huge circus tent, found six employees who had circus experience to pitch it, and put it to use as a church, a movie theater, and an auditorium. She even managed to persuade the military to plant lawns around the barracks.

Perhaps as important as any of this was the personal encouragement and understanding that Maris supplied. She spoke to new arrivals about the importance of their work (Williams 1993), and, according to General Groves, "served as a kind of wailing wall, maintaining day and evening office hours during which the women could come to her with their complaints and troubles and sorrows, knowing they would find comfort and sound advice" (Groves 1962, p. 91). Her effort proved critical to morale and the smooth operation of the Hanford site.

Maris worked very long days and at least part of most nights. Because she didn't have time to drive the forty miles to her barracks, the Hanford management gave her the top floor of one of the three houses still standing on the Hanford town site, which she turned into a two-bedroom apartment. A widow, she shared the apartment with two female friends and her teenage daughter, who came to visit from college. In spite of her long hours and the fact that the rest of the building was used as a day-care center, Maris managed to establish a home. When she asked for a lawn, the Army handed her a sack of grass seed.

In September 1944, Maris returned to her university job, but her work shaped life at Hanford during the rest of the war and beyond. The scout troop and the library she founded are still in operation.

## THE WACS

In the wartime environment, technically trained personnel were constantly in short supply. Mounting pressure to produce the bomb allowed Manhattan Project leaders to use uniformed military personnel. The project was assigned members of the military police; the Provisional Engineer Detachment, a service unit that dealt with such day-to-day matters as food and garbage; the technically trained Special Engineer Detachment; the Women's Army Auxiliary Corps (WAACs); and, after it was established in August 1943, the Women's Army Corps (WACs).

WACs were posted to all three major sites. Hanford had a small group: sixteen to twenty-four women assigned to production work. At Los Alamos, in contrast, the WAC detachment numbered 260 by August 1945. They were initially assigned to clerical and service jobs, but several had technical training and moved into research positions. They handled highly classified material. Twenty WACs worked at the hospital in Los Alamos (Truslow 1991).

Like the MPs and military engineers assigned to the Manhattan Project, the WACs responded to the challenges with varying degrees of enthusiasm. All of them were volunteers, and some, such as Sylvia Galuskin, had volunteered for overseas duty and were frustrated to find themselves surrounded by a bunch of university professors in the middle of nowhere. Galuskin wound up processing work orders at Los Alamos and then worked as a secretary (Schwartz 1991). Mary J. Whiting, Elsie Pierce, and Marjorie Powell Shopp drove in the motor pool, and Florence Pachter Schulkin managed the fund kept to allow civilians to cash checks (Whiting 1991; Shopp 1991; Schulkin 1991; Pierce 1991).

Pierce had joined the Army to match her older brother, and broke an engagement when she was assigned to Los Alamos. Irma Dowd Maeder was sent to Los Alamos along with her husband, Elmo, an Army engineer. She was in charge of ordering supplies, such as candy, beer, and magazines for the PX (Maeder 1991). Other WACs dealt with the endless flow of paper. These included Louise Dumas; Frances E. Steele in finance and intelligence; Iris Bell in the security office; and Mary M. Shanklin, a telephone operator (Dumas 1991; Steele 1991; Bell 1991; Shanklin 1991).

At Hanford, Hope Sloan Amacker worked in the Transportation Department and then for Military Intelligence, monitoring phone calls into and out of the site. She then transferred to Public Affairs, a quiet job but one to which, she was told, she might be called in unexpectedly. On August 6, 1945, she got an emergency call at 7:30 A.M., and for the next two weeks the office was anything but quiet. A native of Middletown, Ohio, Amacker had trained as a telephone operator before enlisting. She thought that Hanford sounded glamorous, with ski slopes nearby—until she arrived. She is also remembered for winning a local beauty contest, which she entered wearing her uniform rather than the standard evening dress (Williams 1993, Sorenson 1995).

Gladys S. Burnett studied education at a women's college in Milwaukee and at Eau Claire State Teachers College. She taught primary school for a year, then married and stopped teaching, though she had no children. Seventeen years later she and her husband separated, and Burnett, who wanted to leave Eau Claire, joined the WAACs, switching to the WACs after they were created. She had learned to operate a switchboard in civilian life and was snatched up by the Manhattan Project to work the Quebec Conference, where Roosevelt and Churchill established guidelines for British participation in the construction of the atomic bomb. Her next assignment was Los Alamos, where she worked in the technical area as a switchboard operator (Burnett 1991).

The WACs' barracks at Los Alamos stood across the street from the men's barracks. Male callers walked up the steps to the door and found that they could see the entire interior of the building. Jane Heydorn suggested a ban on callers, but was laughed out of the idea. She then crusaded for the WACs to stop displaying their tailored, olive-drab underwear on clotheslines, on the grounds that it corrupted the men. (The WAC barracks were also home to Mac, a large boxer owned by

Sergeant Miriam White. Mac, named after General Douglas MacArthur, had free run of the base, including the technical areas.)

In a closed city with ten men for every woman, the WACs were popular figures. Jerry Stone reports receiving three marriage proposals during her work at the PX, none of which she took seriously. By 1945, so many soldiers and WACs had married that a special apartment was set aside for their use, in rotating two-week intervals. Pregnant WACs were discharged immediately (Roensch 1993).

Military pay was much lower than civilian. WACs had to live in barracks and were subject to military drill and, except on Saturday night, bed checks. Some of the civilians tended to treat uniformed military personnel as second-class citizens, wrapped up in red tape and a bit stupid. Fortunately, most of the technical personnel treated the military staff as colleagues and recognized their contributions. After Hiroshima, the WACs generally took pride in the contributions they had made to winning the war.

The WACs came from a wide variety of backgrounds—and had a wide variety of futures. Captain Arlene Schiedenhein commanded all the WACs working on the Manhattan Project and shuttled among project sites. After the war, she received the Army's Legion of Merit and a *Mademoiselle* Merit Award (1946), an honor she shared with Leona Marshall (*Mademoiselle* 1946). In contrast, one WAC sergeant was arrested when the war ended. She was a tiny, tough, and very popular woman who worked as a truck driver, shuttling supplies from the railroad at Lamy to Los Alamos. While in town, she would buy tequila at $3 a fifth and sell it on the base for $15. Officially, the military personnel at Los Alamos drank nothing stronger than 3.2 percent beer, so the sergeant's service was both popular and profitable. She also shot craps and played a good hand of poker. Rumor says she used to win most of the pay coming into the motor pool. At the end of the war, the government ceased to turn a blind eye to her activities and had her arrested. She spent only one night in jail before former GIs bailed her out, but her military career ended abruptly.

Norma Gross, a New Yorker and a first-generation American Jew, joined the Army because it agreed to accept her as an enlistee as soon as her husband, a physicist who had been working on electronics for the radar project at MIT, was shipped from New Jersey to his military posting. The Navy offered Gross an officer's berth, but wanted her immediately. Gross was older than most enlistees, and there were other differences as well. When she asked, in sincere ignorance, "What does a bedbug look like?" her teenage barracks mates thought she was putting on airs.

After basic training and a spell at Fort Sill, Oklahoma, with "lovely quarters, good meals, and lots of freedom," Gross found herself dumped with a group of about eight WACs at the Lamy train station, waiting for the trucks that would carry them to the hectic world of Los Alamos. Once she arrived, a new and unusual problem arose. Gross's husband was an officer and she was enlisted, so they could fraternize only with the permission of her commanding officer. Gross remembers that civilian friends were very kind about letting the young couple use their houses. Eventually the commanding officer decided that Gross was having too much fun and withdrew permission for the couple to be seen together (Gross 1997).

Pat McAndrew, a WAC sergeant, arrived at 109 East Palace Street with two other WACs in tow and her secret orders pinned to her brassiere. Since the East Palace office was closed on Sunday, the three women had some difficulty getting to the mesa. McAndrew managed eighty-five people and kept the records for all the money that went through the laboratory. The relaxed Los Alamos style came as a surprise to her. In her words, "The first time I met my CO, she was seated at her desk dressed in this red pajama-like outfit. I was in my neatly pressed uniform, and I was taken back" (Hogsett 1980, p. 11).

If military discipline at Los Alamos left much to be desired, the accounting practices were even stranger. As McAndrew tells it:

> I was always worried because my instruction always came by telephone. I'd get a call and this voice would tell me to take the money, for whatever it was for, and meet a certain captain in Tesuque [a pueblo near Los Alamos]. I'd hand over the money and he would hand me a slip that said, "received from A. Karam—$400,000," that's all. My name was Karam back in those days. Like I said, I was very military and I'd take things too seriously. Here'd come this captain, hat cocked, and chewing a big wad of gum. I was always reluctant to hand such sums—and it was always cash—over to someone like that. My instructions were always over the phone. I had nothing in writing, but it always worked and I never got into trouble." (Hogsett 1980, p. 11)

McAndrew also censored letters. She frequently returned them to their authors for postage if they were personal and for corrections in grammar or spelling if they were official. She knew nearly everything that went on in the secret world of the Manhattan Project. One of her subordinates, Dave Heimbach, describes her as "tough, mentally and physically, and she was extremely fair" (Hogsett 1980, p. 11).

Many workers, of course, resented censorship. Norma Gross complained when a letter from her husband arrived full of holes. Her husband's brother, a physicist working in industry in Connecticut, had written to her husband describing the departure of numerous friends for Clinton and New Mexico, and the censors had cut out the names of all the towns to which the men were traveling. Gross complained that it made no sense because she already knew what was happening at each town (Gross 1997).

## THE CIVILIANS

Probably the largest single group of female workers on the Manhattan Project was made up of the local women recruited by Tennessee Eastman to operate the enormous calutrons, known as racetracks, at Clinton Engineer Works. The calutrons were giant mass spectrometers that used electromagnets to focus a beam of uranium ions into a semicircular path. The less massive uranium-235 ions—the ones that were needed for fission—followed a path with a slightly smaller radius than that of the uranium-238 ions. The technique had been tested by Ernest Lawrence in his Berkeley laboratory, and he argued successfully to have two stages of electromagnetic-separation calutrons constructed at Oak Ridge. In typical Manhattan Project manner, the contractors broke ground on February 18, 1943, before the final design for the calutrons had been completed.

A calutron unit consisted of two tanks from which the air had been evacuated, set between the poles of a large, square electromagnet. The magnets were arranged in an oval of ninety-six tanks and forty-eight magnets for the alpha, or first-stage, separation process. Second-stage (beta) calutrons were half as large and arranged in rectangles with thirty-six units (Groueff 1967, p. 268). Because copper was needed for the war, Groves borrowed 13,450 short tons of silver, worth more than $300 million, from the Treasury for magnet windings.

The calutrons themselves were located on the second floor of racetrack buildings, while the pumps for circulating oil in the magnets and those for maintaining vacuum in the tanks were on the ground floor. Each tank weighed fourteen tons, but the magnetic fields could shift them out of alignment by as much as three inches, so they had to be

## Other Women of the Manhattan Project 169

welded to the floor. The magnets underwent one complete redesign, and systems had to be devised to clean both the windings and the cooling oil so that the magnets would not short out (Rhodes 1986). At least 90 percent of the uranium processed by the calutrons remained in the system, so the entire setup had to be shut down about every ten days to recover the precious isotope (Groueff 1967).

The calutrons operated around the clock, seven days a week. By August 1943, Tennessee Eastman had realized that it would need to hire some 13,500 operators. The first calutrons at Berkeley had been operated by Lawrence's team of physicists and graduate students, but hiring 10,000 physicists was impossible. Even if they could have been persuaded to concentrate on the exacting but boring job of monitoring calutrons, there simply *weren't* enough physicists. Instead, Eastman hired local women and trained them to watch certain meters and to keep the readings within specific limits. They didn't need to know what the meters measured or what the knobs and levers actually controlled.

"Our wives all drive automobiles," the argument went. "Do they have to know about the principles of combustion? They've never seen what's under the hood, but they are perfectly able to operate an engine." Nevertheless, the idea that a woman could perform electromagnetic isotope-separation was a bitter revelation for some scientists (Groueff 1967). Only when the newly trained operators outproduced scientists operating calutrons were all the doubters convinced.

Eastern Tennessee proved a fruitful recruiting ground for the project. The major regional industry was coal mining, and women worked either as waitresses or in stores. The relatively high government wages offered by the Manhattan Project—75 cents an hour for a six-day week (Groueff 1967)—brought a flood of applicants. Some answered ads in local papers; others learned about the jobs from recruiters or by word of mouth. Most of the women were particularly eager to contribute to the war effort. They had no idea of what the project was, but they had been told that it was essential to winning the war.

Creola Green McCamey, a coal miner's daughter, graduated from high school in eastern Tennessee. She had cousins working on the project and could stay at her aunt's house. As soon as she turned eighteen, she went to the employment office in Oak Ridge and was hired. For three weeks she reported to a training center, where she was shown drawings of knobs, levers, and meters and told what she would be doing. After her security clearance came through, she reported for work at the Y-12 plant.

Workers passed through a guarded fence and began their shifts in the change house, where they donned color-coded uniforms. McCamey's was blue. The workers then reported to their "cubicles," which were machines covered with controls. The calutrons' huge magnets caused unusual problems: They could pull hairpins out of hair and tools out of hands, and when they captured something, it took strong men to pull it free (Groueff 1967). A magnet once had to be shut down before a three-foot length of pipe could be removed.

The operators checked the meters and adjusted the knobs and levers to keep the readings within set limits. If they had trouble, they called the "floor lady," who was in charge of five or six machines. She, in turn, could call on other technicians for help. If the machines were working well, an eight-hour shift got boring, but things could get hectic if there were problems.

McCamey worked rotating shifts: seven days working from 7 A.M. to 3 P.M., a day off, seven days working 3 P.M. to 11 P.M., a day off, and seven days working 11 P.M. to 7 A.M., followed by four or five days off. Each shift had a one-hour lunch break. In spite of the hard work, McCamey found the job fun. The food was good, the buses ran on time, and the workers considered themselves a team, bound by patriotism. As McCamey put it, "We were young and full of vinegar." She had moved in with a friend whose husband was overseas and who had three small children, and McCamey's grandmother kept house for them (McCamey 1996).

Pat Patterson lived with her family and commuted to Oak Ridge. Like McCamey, she had just graduated from high school and loved the independence of earning good money. Helen Hall, a recent high-school graduate from Murfreesboro, Tennessee, was recruited by a customer in the drugstore where she worked. She and a friend cashed their war bonds and caught a bus to Knoxville, where they were hired for the project. They were assigned a double room in one of the new dormitories at the site. They enjoyed the excitement of being away from home and part of a team working to win the war (Groueff 1967).

McCamey remembers hearing about the Hiroshima bomb in a bus terminal. People were out of their heads with excitement, waving newspapers and yelling. She still has the Army/Navy E (for excellence) pin that she and the other operators were awarded.

Many of the civilian women working on the Manhattan Project were clerks or secretaries. Ida M. Coveyou married in September 1943 and joined her husband at Oak Ridge in December. She worked in the Accounting Department, determining how much money should be deducted from monthly paychecks for rent. She took advantage of night classes to continue work on her undergraduate degree (Coveyou 1991).

Joanne Stern Gailar was a nineteen-year-old living in New Orleans when she became involved with the Manhattan Project through her fiancé. "You can't, in your wildest dreams, imagine how secret this mess is," he wrote to her. "Everything is marked 'confidential' and every department and sub-department is a unit in itself, and when the product from the department in which I work goes through the wall into the next, it is gone—nobody in my gang knows where it goes or what happens to it. Likewise, the stuff that comes into the department I work in, nobody knows where it comes from, what it is for, who made it or how. All we know is our own specific little research problem . . . " (Gailar 1990). Gailar moved to Oak Ridge, and in 1945 worked as a clerk at K-25, which after the war became the Oak Ridge gaseous-diffusion plant. Twenty years later, she returned to work at Oak Ridge as a defense analyst specializing in the Soviet Union. She worked with the physicist Eugene Wigner, with whom she was the coauthor of several publications.

Many women in administrative work married colleagues on the Manhattan Project. Sonia Weiner, secretary to the assistant director of the Met Lab's Chemistry Division, married Joseph Katz, a chemist, in the fall of 1944 (Kathren et al. 1994). He became a senior chemist at Argonne National Laboratory after the war, and she became a prominent artist in Chicago.

Mitzi Mars Butcher, a farm girl from rural Montana, was transferred to Hanford from a job at a Remington Arms munitions plant in Salt Lake City. She remembers being shocked by the rough-looking men she met when she and four girlfriends stepped off the bus in Pasco. She was assigned to the Employment Office, where she worked twelve-hour days, six days a week (Williams 1993).

Geneva Owen Hammer dropped out of a Missouri high school to follow her parents to Hanford. Her father drove buses to the B reactor and her mother sold cigarettes and sundries in a Hanford beer hall. Hammer worked in a cafeteria as a coffee girl until she quit to take a job in a shipyard (Williams 1993).

Mary Lane received a bachelor's degree from North Texas State University in speech, drama, and physical education. In 1943 she went to an employment agency in Cross Plains, Texas, which found her a position at Los Alamos. She reported to 109 East Palace, where she met both Dorothy McKibbin and the biggest German shepherd she had ever seen. Already terrified by the dog, she was horrified by the road to Los Alamos, which was edged by sheer cliffs and so narrow that trucks had to back around some of the tighter curves.

Lane started work as a clerk and typist for the stockroom. To get raises, she changed jobs several times, becoming an accountant, a pasting-machine operator, and a supervisor in the electronics stockroom. Finally she was assigned to the Cyclotron Group as a secretary; her job was to keep track of the precious metals used as targets for the cyclotron beam. The group asked her to work experimental shifts, but she managed to wander close enough to the beam to get a dose of radiation that made her ineligible for experimental work (Lane 1991).

Many Los Alamos wives were drafted into nontechnical work. Laura Fermi recalls that she was paid the minimum wage for three hours a day. She had a blue badge, which meant that she could enter the Technical Area, though she was not cleared for secret work. She writes of her experience on being hired for clerical work in the Health Division: "My boss, Dr. Louis Hempelmann, was very conscious of [the blue badge] the day he hired me. He was a tall, willowy young man with a shock of blond hair on his forehead. I was then thirty-seven years old and probably the oldest person he had ever employed. He was my first paying boss (I had done only volunteer work so far), and we both acted shy. His embarrassment showed in his easy blushing, which made him look little older than a schoolboy; mine in verbiage and too many questions" (Fermi 1954, p. 228).

Many women who did not work directly on the Manhattan Project did other work in the hastily constructed communities around the sites. Ellen L. Tinsley and Cora F. Cooper followed their husbands to Los Alamos. In order to get good housing and household help, both husband and wife had to be employed, so Tinsley and Cooper taught in the growing Los Alamos school system. The teachers struggled with a wide variety of students in the closed society of the Hill, and worked with mothers to provide recreation for the children (C. Cooper 1991; Tinsley 1991). Other teachers included Alice Kimball Smith, who agreed to manage all social science; Dorothy Hillhouse, who ran a Brownie Scout troop and organized a square-dance club in her free time (Manley 1990); Jane Wilson; a chemist named Mrs. Long, who had teaching experience; Betty Inglis, who was experienced in teaching mathematics and also worked on computations in the Theoretical Division of the laboratory; and Jean Parks Nereson (Brode 1980).

Alice Kimball Smith, wife of Cyril Stanley Smith, a Los Alamos metallurgist, served on the Town Council and taught history in the high school. In 1946, when her husband went to the University of Chicago to become the first director of the Institute for the Study of Metals, she worked as an assistant editor of the newly established *Bulletin of the Atomic Scientists*. She wrote an insightful historical study of the Manhattan Project scientists' postwar campaign for civilian and international control of atomic energy, and eventually, when her husband joined the faculty of MIT, became dean of the Bunting Institute at Radcliffe College (Hoddeson et al. 1993; Smith 1988; Manley 1990).

Many women played prominent and important roles in technical writing and information handling. Edith C. Truslow wrote the official history of Los Alamos with Ralph Smith, along with other works about the project. Bernice Morgan Burkhardt also worked with Smith in the Document Division. She joined the project because it seemed the patriotic thing to do, and because the job paid better than other jobs for writers in the Santa Fe area. She later became an award-winning freelance journalist and the mother of three (Burkhardt 1991). Wilma Chiotti ran the library at Iowa State College, which supplied chemically pure materials needed for the project. Chiotti was responsible for obtaining all classified publications (Svec 1995).

Marcella D. Lazarus transferred to the Manhattan Project from the intelligence branch of the Army Corps of Engineers. As a recruiter for the Met Lab, she traveled the country persuading scientists to join the project and talking their universities into letting them go. She also dealt with Selective Service Boards, making sure that critical scientists would not be drafted—or, if that failed, arranging to have the scientists assigned to the Special Engineer Detachment and posted to the appropriate project sites. She worked long hours, like other project employees, and got most of her sleep in the lower berths of trains (Lazarus 1993).

E. Ann Jensen started at Los Alamos and continued her career at Lawrence Livermore Laboratory (Jensen 1991). Ida Rizzoli worked in the

Accountable Property Department at Los Alamos with Anna Mary Sites and Isabelle Trujillo. The office kept track of the enormous quantities of equipment ordered by the growing laboratory. When Rizzoli joined her husband at Los Alamos, she had to quit an official job at the University of California and be rehired by the civil service, but she had no idea for whom she was working. There were no signs on doors, and she was the only civilian in an office full of young GIs (Rizzoli 1991).

Dorothy Myers, having taken courses in bookkeeping and business math in high school, was hired as a secretary in the Procurement Division at Los Alamos. She found herself in charge of the stationery stock, which supplied all the laboratory offices in the Tech Area. She claims that she and her coworkers walked about ten miles a day in the course of their work and had to be issued extra shoe-ration stamps. On October 31, 1945, Dana Mitchell, assistant director of Los Alamos, wrote her a letter of appreciation. His words summarize the contributions of many of the female workers on the Manhattan Project:

> When you are leaving Project Y, I will not be here to tell you how much I personally appreciate the service that you have rendered in the multitudinous activities of the Procurement Division. As you know, the service and supply for our combat units was more important in this war than any heretofore. Most of this work was far from the fighting lines and had little of the glory or excitement related to actual combat. Nevertheless, it was of extreme importance and every man who served well has the right to be proud.
>
> In the same way, the Procurement Group at Project Y was intimately related to and of essential importance in the development, the final production and the delivery of the atomic bombs.
>
> Whether or not our success in releasing atomic energy will be good or bad for mankind, only the future can tell. In the meantime, however, we know that our success in producing this bomb so soon, resulted in the saving of untold lives by precipitating the cessation of combat.
>
> I, therefore, consider it a privilege to compliment you and I hope that you have the personal satisfaction of being proud that you did your best in the work at Los Alamos. (Myers 1991)

Women brought new attitudes to their demanding jobs on the Manhattan Project. *(Richard E. Donnell, from The Home Front)*

WACs drill at Los Alamos. *(National Archives, RG 434-OR, Box 14, Notebook 42, MED 348)*

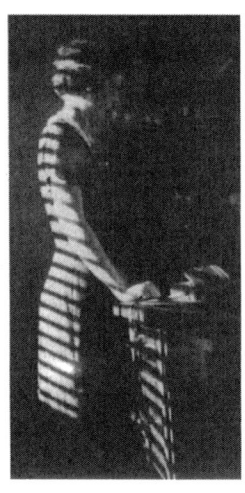

Nathalie Baumbach at work in a laboratory at UCLA. *(Courtesy of Dr. Glenn T. Seaborg)*

Manhattan Project husband and wife team, Dieter and Frances Kurath, shown in Chicago. *(Dieter Kurath, private collection)*

Eleanor Ewing marries Richard Ehrlich on July 7, 1945, in a typical Los Alamos ceremony. *(Eleanor Ewing Ehrlich, private collection)*

Kay Florin outside the New Chemistry Building at the Met Lab. *(Argonne National Laboratory, OLD-NEG #1-1001)*

Oak Ridge workers line up outside the Jefferson Recreation Hall to buy rationed cigarettes. *(Photograph by James E. Westcott, National Archives, RG 434-OR, Box 14, Notebook 43, PRO-469)*

Workers line up in "Clock Alley" at the Clinton site. *(Photograph by James E. Westcott, National Archives, RG 434-OR, Box 14, Notebook 43, PRO-763-2)*

African Americans worked on the Manhattan Project but were segregated, as illustrated by these Oak Ridge privies labeled 'white' and 'colored' outside the X-10 plant. *(National Archives, RG 434-OR. Bpx 14. Notebook 43, X-10-14)*

Segregated facilities at Oak Ridge included this Colored Recreation Hall. *(National Archives, RG 434-OR, Box 23, Notebook 70, PRO 1310-1)*

WACs relax in the day room in their dormitory at Oak Ridge. *(DOE photograph by James E. Westcott, Neg. #1679-2)*

A bowling team organized at Hanford. *(Department of Energy photograph, Hanford National Laboratory, D2083)*

Met Lab workers relax at the Indiana Dunes. *(Courtesy of Dr. Glenn T. Seaborg)*

A street scene in the town of Hanford. *(National Archives, RG 434-OR, Box 14, Notebook 41, A-56)*

Workers in the cafeteria at Hanford. All the food you could eat for three meals a day was one of the recruiting promises that lured construction workers to the remote site. *(National Archives, RG 434-OR, Box 14, Notebook 41, MED 376)*

Shift change at the Y-12 plant in Oak Ridge. *(Department of Energy Photograph by James E. Westcott, neg. #PRO 936)*

Workers at Hanford each contributed a day's pay to purchase this B-17 Flying Fortress, which flew 47 combat missions from July 1944 to the end of the war. The woman in the delegation appears to be Buena Maris, although she is not officially identified. *(National Archives, RG 434-OR, Box 14, Notebook 41, MED 375)*

# mlle merit awards

### for these ten, 1945 meant signal achievement

| | |
|---|---|
| dr. leona woods marshall | physicist |
| capt. arlene schiedenhein | WAC |
| barbara bel geddes | actress |
| emily wilkens | designer |
| patricia van delden | resistance organizer |
| gwendolyn brooks | poet |
| ruth m. leach | business executive |
| susie reed | ballad singer |
| blanche thebom | opera singer |
| u.s.s. missouri | fleet glamour girl |

**barbara bel geddes**
turned New York on its ear this past season. As the senator's daughter in the sociological drama, *Deep Are the Roots*, she battered down two of the toughest barriers in her career: being known as the daughter of a famous man (industrial designer Norman Bel Geddes), being trapped in ingenue insipidities. Fevered acclaim for her sensitive performance placed her as a separate theatre personality; boosted her up to stardom

**capt. arlene schiedenhein**
was commanding officer for WACs at all installations concerned with the atomic bomb. Her job: handpicking her highly trained personnel; shuttling between four hidden plants in widely separated states; bolstering welfare and morale as WACs served long shifts preparing, filing, handling all top secrets. For her outstanding services in connection with the Manhattan Project, she won the Army's Legion of Merit

**dr. leona woods marshall**
represents the continued progress of American women in science. With a Ph.D. for her research in spectroscopy at the U. of Chicago, she was one of a group of physicists, directed by Prof. Enrico Fermi, who worked three years on the atomic bomb project. Unexpected award for Leona—romance: on this same project she met and married Dr. John Marshall. Together, they will return to experimental physics

**emily wilkens**
designs gay clothes for young juniors. Four years ago, creation of special fashions for teensters was an orphan industry, formless as their Sloppy Joe sweaters. Emily flattened the tummy, camouflaged the no-waistline and lankiness, let her pencil do what it had wanted to do in her former fashion illustrating job. First winner of the Fashion Critics' Award, she has her own Emily Wilkens Shops in country-wide stores

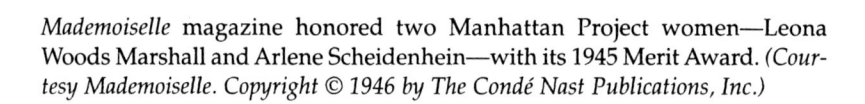

*Mademoiselle* magazine honored two Manhattan Project women—Leona Woods Marshall and Arlene Scheidenhein—with its 1945 Merit Award. *(Courtesy Mademoiselle. Copyright © 1946 by The Condé Nast Publications, Inc.)*

# 9  After the War

WHEN JAPAN SURRENDERED in the wake of the bombings of Hiroshima and Nagasaki, the American public credited the atomic bomb with removing the need for an Allied invasion of Japan that would have cost hundreds of thousands of lives on both sides. Manhattan Project leaders became heroes, and many project personnel learned for the first time exactly what they had been creating.

Before Hiroshima, a few serious thinkers had recognized the problems inherent in the awesome destructive power of the atomic bomb. The techniques of its construction could not remain secret forever; should scientists demand a role in controlling both the use of atomic weapons and the dissemination of the knowledge needed to produce them?

The discussion started at the Met Lab. In Chicago, work pressure had decreased once plutonium and enriched uranium production got under way at Hanford and Clinton. A standard forty-eight-hour workweek left time for thinking. While the war continued, constructing an atomic bomb had seemed a necessity, but in peacetime, scientists began to question the morality of continuing to produce such weapons.

Another consequence of peace was massive demobilization. Families throughout the country welcomed back fathers, husbands, fiancés, sons, and brothers. Economists worried about the effect that the flood of returning soldiers would have on the job market. Fearing another Depression, they invented a program encouraging veterans to seek higher education. The GI Bill channeled veterans into America's universities, but tens of thousands of soldiers still needed work.

The public-relations mechanisms that had actively encouraged women to work for the war effort now appealed to them to give their jobs to returning soldiers. The popular press lauded the woman who cooked and cleaned for her family and bore children for her husband. Commercials showed women vacuuming their homes in fresh housedresses and high heels. Rosie the Riveter, poster girl of the war, was definitely out of fashion.

Many women did not question the returning veterans' right to jobs. They stepped aside willingly, returning to their families or starting new ones. But for other women, being forced out of their jobs entailed wrenching change and consigned them to "women's work," with its accompanying reductions in responsibility and pay. They had shown that they could do scientific work, and they were frustrated that they were not allowed to continue (Wise and Wise 1994).

Life at the laboratories of the Manhattan Project changed abruptly. The tremendous pressure to produce a working weapon was suddenly gone. The communities became more open. Los Alamos employees could visit their families for the first time in years. Many of the scientists were eager to return to their university and industrial jobs. The WACs and SEDs were demobilized, and the industries that had run the labs turned eagerly to profit-making peacetime operations. Congress and the executive branch debated whether to continue the wartime laboratories and how to support or disband the massive research apparatus they had constructed.

## ATTITUDES TOWARD WORK ON THE ATOMIC BOMB

Almost without exception, the women who worked on the Manhattan Project recall the experience as a major event in their lives. The project moved at a breakneck pace, making discoveries under enormous pressure. Young staff members forged close lifetime friendships. And the Manhattan Project parties were, by all accounts, nearly as extraordinary as the work.

"When we were employed we were told to ask no questions, and we didn't—much," wrote Dorothy McKibbin. "We worked with pride. We sensed the excitement and suspense of the Project, for the intensity of the people coming through the office was contagious. Working at 109 was more than just a job. It was an exciting experience. Our office served as the entrance to one of the most significant undertakings of the war or, indeed, of the twentieth century" (McKibben 1988, p. 27).

Los Alamos, in Charlotte Serber's words, "was anything but conventional. Everything was new and different and frantic. But we did a job and evidently did it well. History seems to have been written at Los Alamos, and I think the working wives can claim a small share of the writing" (C. Serber 1988, p. 71).

Hope Amacker, a WAC sergeant, writes that her work at Hanford "was the greatest experience I could have had outside of going into combat. It was a lifetime experience. We were made to feel like we were contributing to the war effort. I never felt I was just a peon" (Williams 1993).

For many Manhattan Project women, a sense of responsibility for the weapon they had helped to create accompanied the pride they took in their work, but most, like most Americans in general, seem to have felt that the creation of the atomic bomb had been necessary. For instance, Leona Marshall, who worked on the original atomic pile at the University of Chicago, did not regret her contribution to the development the first nuclear weapon. "As awful as that bomb was," her son recalled, "she said it saved a lot of lives, with the invasion casualties estimated [at] at least a half-million people" (Folkart 1986).

Some women, though proud of their contributions, simply did not talk about their work on the Manhattan Project because their children grew up in the turbulent '60s and '70s. "We didn't talk because the kids were horrified by the bomb," said Norma Gross. "They thought anybody who worked on it was a murderer, and it made no sense trying to explain about the threat."

Melba Robson, who worked as a biomedical technician at the Met Lab, never forgot her fears that the first atomic test would ignite an uncontrolled fusion reaction in the atmosphere. While in Oak Ridge, she married Arthur Robson, an engineer from Kodak, following him through a variety of jobs after the war. She eventually accepted a position as associate professor at the University of Illinois School of Allied Arts in the Medical Sciences (A. Robson 1995; Watson 1995). Her son writes, "In 1959, the air-raid sirens started blaring. While we were packing for the move into the bomb shelter, we learned from the radio that the Chicago White Sox had won the pennant. Melba knew that she had helped to light the fuse that would destroy the world" (A. Robson 1995).

## WOMEN AND THE MOVEMENT FOR CONTROL OF ATOMIC ENERGY

By early June 1945, scientists at the Met Lab had expressed grave concerns about the future use of atomic weapons. Arthur Compton appointed a formal committee, headed by James Franck, to discuss these issues. On June 11, the committee submitted a report to the Secretary of War, Henry

Stimson, recommending that the bomb be used not in combat but in a demonstration that would persuade the Japanese to surrender. The Franck Report also foresaw the development of an arms race and urged international control of atomic energy (Smith 1965, Appendix B).

Compton faithfully reported the concerns of his scientific staff to the Interim Committee, a high-level government group charged with deciding how to use the first atomic bomb. Various scientists, notably Leo Szilard, attempted to contact high government officials directly. The scientists' views, however, counted for little. For one thing, they were far from unanimous. In an informal poll that Compton took of his staff, 13 percent of the respondents favored not using the atomic bomb at all; 46 percent favored its use on a military target in Japan, followed by a demand for surrender; 26 percent wanted a demonstration in this country; and 15 percent considered the decision the prerogative of the military (Smith 1965).

At Los Alamos, Robert Oppenheimer felt strongly that political action was inappropriate for scientists engaged in a struggle to produce a working weapon. His influence, together with the tremendous work pressure, stifled discussion (Smith 1965). Many of the Los Alamos staff were stunned when the Trinity test showed them the power they had unleashed, but their efforts to produce a working implosion bomb seem to have exhausted them. Scientists at Oak Ridge were not even officially informed that the Trinity test had been successful.

Only three weeks elapsed between the Trinity test on July 16 and the first use of the bomb. The news of the destruction of Hiroshima burst over the laboratories on August 6, 1945. Until then, many workers had had little idea of what they were developing. Hope Sloan Amacker, who worked in public affairs at Hanford, had been told to expect an emergency call in early August, but she did not know what she would be publicizing until August 6.

Mrs. Fritz Matthias, wife of the military commander at Hanford, recalls that on August 6, her husband advised her to keep the radio on, but refused to tell her why. She was grinding ham for a party when she heard about the bombing of Hiroshima. She later recalled her shock and how, with tears in her eyes, she phoned her husband for confirmation. The day passed in a blur of newspapermen and ended with a party. "I couldn't help but believe that God, wearying of this long and tortuous war, had finally, reluctantly, given us this terrible weapon with which to end it," she said (Van Ardsol 1958, pp. 74–79).

At Los Alamos, the scientists expected some news of the atomic bomb, but did not know what it would be or when it would come. Lilli Hornig recalls relatively little discussion of the ethics of using the bomb, thought she did sign a petition urging that the first one be used on an unpopulated island as a demonstration. Laura Fermi describes the announcement of the Hiroshima explosion thus: "In Los Alamos, the paging system announced the news in the Tech Area, and the men were stunned. A blow is no less painful for being expected" (Fermi 1954, p. 244). Phyllis Fisher, wife of Leon Fisher, a physicist from Berkeley, realized that the colored cylinders her husband had brought home as children's toys were the casings from parts for the bomb. She had strung them together to make Christmas ornaments (Fisher 1985).

Most women reacted first with surprise and then with relief, since it was clear that the war was about to end. Like the rest of the country, they looked forward to peace and were proud of their contributions (J. Howes). From the weary bomb-assembly teams at Los Alamos to the people on buses at Hanford who shouted the previously forbidden word "plutonium," the predominant reaction of Manhattan Project personnel who suddenly understood the purpose of their work was joy in its success and the release from wartime pressure.

In September 1945, Margaret Smith Stahl, a social scientist, surveyed 450 chemists and physicists selected at random from membership lists of the American Chemical Society and the American Physical Society—a group heavy with European immigrants. Nearly half of them responded, and 97 percent of those felt that the government had been right to sponsor the development of atomic weapons. More than two-thirds of the respondents thought that the United States had been right to use the bomb on Japanese cities without warning even if the war could have been won without it; 20.8 percent would have preferred its use on cities after a warning (Smith 1965). The atomic bomb was a technological triumph, and its makers rejoiced in their success. Women were not surveyed separately, but our interviews indicate that their reactions were similar to those of the men.

The Manhattan Project labs joined the rest of the country in celebrating the end of the war with parades, snake dances, and drunken parties. The gaiety covered the beginnings of serious concern about the future of atomic energy in general and the future of the labs in particular. Graphic reports of the destruction at Hiroshima and Nagasaki increased the scientists' concern. The shift from the struggle to produce a weapon that would end the war to the consideration of the implications of nuclear weapons proved emotionally wrenching. Laura Fermi found that her husband and his colleagues "... assumed for themselves the responsibility of Hiroshima and Nagasaki, for the evils that atomic power might cause anywhere, at any time" (Fermi 1954, p. 241).

On August 23, a committee of Los Alamos scientists called a meeting of staff members interested in forming an organization aimed at maintaining a scientific voice in the use of atomic energy. Its purpose was described in a mimeographed flier:

> Many people have expressed a desire to form an organization of progressive scientists which has as its primary object to see that the scientific and technological advancements for which they are responsible are used in the best interests of humanity.
>
> Most scientists on this project feel strongly their responsibility for the proper use of scientific knowledge. At present, recommendations for the future of this project and of atomic power are being made. It would be the immediate purpose of this society to examine our own views on these questions and take suitable action. However, the future will hold more problems and scientists will feel the need of a more general organization to express their views. (J. Hinton 1990)

A meeting took place on Thursday evening, August 30, and a committee was formed. Female members of the technical staff were invited to this meeting, and several became actively involved in the work of the committee. Women who were not on the technical staff were not asked to the August meeting, but in October a second notice was sent out:

> TO ALL WOMEN LIVING AT LOS ALAMOS:
>
> What the President and Congress decide to do about the atomic bomb "secret" is *our* business, too.
>
> All over the country people are thinking and talking about the atomic bomb and what it means to the future of America and the world. Much is being said or written that is based on wrong information or misunderstanding of the facts. We at Los Alamos have every reason to be particularly anxious that people know the truth about atomic energy and its possible use for good or evil—that it can never be the exclusive property of this or any other country.
>
> The Los Alamos Association of Scientists is an organization of men working here who are trying, by every possible means, to get the facts to the American public. This is an enormous job, and <u>everyone</u> who is interested will have to work if we are to accomplish anything in the short time before the final decisions will be made.

If you are willing to help, come to a Women's Meeting in the High School Library Monday, October 8, at 8:15 P.M., to discuss what we can do and how we should go about it. Mr Higginbotham from the men's committee will tell us what work it feels is most urgent and how we can share in it. (J. Howes)

This notice clearly reflects some women's perceptions that men should lead the movement for international control of atomic energy, a view that prevailed in spite of the leadership role traditionally played by women's organizations in the peace movement (Washburn 1993). The women's meeting developed into a women's auxiliary led by Kay Manley, which handled such chores as typing and monitoring press reactions to atomic issues. Dorothy McKibbin used her Santa Fe connections to start a program of contacts between Los Alamos scientists and the community (A. K. Smith 1965).

At least some of the female scientists involved in the Manhattan Project played active roles in the movement to internationalize atomic energy. Joan Hinton worked with colleagues at Los Alamos to send samples of the glass-like substance formed from melted sand and equipment in the crater left by the Trinity test to the mayors of the country's largest cities. The samples were accompanied by a message that read in part, "Do you want this to happen to your city?" (J. Hinton 1990) Maria Mayer served as a member of a negotiating team for the Association of Manhattan Project Scientists and was a guest at a dinner for scientists and senators (A. K. Smith 1965).

Hoylande Young was active in the Atomic Scientists of Chicago. Jane Hall joined the scientists' movement in Chicago and continued to work in Los Alamos, where she served on the executive committee of the Association of Los Alamos Scientists and participated in radio broadcasts aimed at educating the public about atomic energy. Kay Way was active in Chicago and later at Oak Ridge, became Director of Publications for the Federation of Atomic Scientists, and coedited the influential collection of essays *One World or None,* published in 1946 (A. K. Smith 1965). In this book, which sold 100,000 copies, leaders of the scientific community, including Einstein and Bohr, pointed out the dangers of a nuclear world (Artna-Cohen et al. 1993).

The most prominent female scientist in the national organizations was Melba Phillips. Born on a farm in rural Indiana, she impressed the distinguished atomic physicist Edward Condon by correcting his answer to a problem in quantum mechanics during summer school at the University of Michigan. Condon helped her enter the graduate program in physics at Berkeley, where she became Robert Oppenheimer's first graduate student. She received her Ph.D. in 1933 and continued postdoctoral work for the next two years. During this time, she and Oppenheimer discovered the Oppenheimer–Phillips reaction, which occurs when a low-energy deuteron passes close to a nucleus and has its neutron–proton bond broken and its neutron stripped off and absorbed by the nucleus. Such reactions can occur in conjunction with nuclear fusion. Phillips moved to Bryn Mawr as a Helen Huff research fellow and spent 1936–37 at the Institute for Advanced Study in Princeton on a Margaret Maltby Fellowship from the American Association of University Women. She worked as an instructor at Connecticut College for a year before securing a position at Brooklyn College (Bonner 1993).

Oppenheimer frequently referred to Phillips in his letters, and it is obvious that they remained friends (Smith and Weiner 1980). Phillips did not work on the Manhattan Project, however, instead becoming a lecturer at the University of Minnesota and spending five months on the radar project. After the war, she returned to Brooklyn College with a part-time appointment at Columbia University (Bonner 1993). She was a leader in organizing the Association of New York Scientists and frequently represented the group at national meetings. In 1946 she became secretary of the newborn Federation of Atomic Scientists, a volunteer position, and worked at the organization's Washington office for three days each week (A. K. Smith 1965).

In October 1952, Phillips was fired by both Brooklyn College and Columbia University for refusing to answer the questions of the Senate's McCarran Committee on her own and her colleagues' political views. Blacklisted for the next five years, she wrote two textbooks, one a classic graduate text on electricity and magnetism, and several encyclopedia articles. In 1957, Edward Condon invited her to Washington University in St. Louis to head a program for high-school physics teachers.

Phillips had a distinguished career in physics education, eventually becoming the first female president of the American Association of Physics Teachers and the first woman to lead a national physics organization. In 1987, Brooklyn College finally acknowledged the wrong it had done Phillips and other faculty members fired during the McCarthy era, but the college never made financial reparation (Bonner 1993).

Nathalie Goldowski also suffered under McCarthyism. She was forced out of the new national laboratories because her Russian and French background could not be traced, and she did not become a U.S. citizen until 1947. She taught in the Physics and Math Departments at Princeton and then went to the experimental Black Mountain College in North Carolina, where she married one of her students. When Black Mountain College closed in 1952, she worked in industry before becoming a professor of physics and ceramics at Alfred University in New York. Over the course of her career she lectured at the Nobel Institute and the Belgian Royal Academy, and she received a medal for a moving picture that she produced on the corrosion of metals (Bernstein 1992; *American Men and Women of Science*).

Always outgoing and outspoken, Goldowski perpetually clashed with administrators. She would greet colleagues whom she had not seen for several days with an enormous hug and an exuberant "I love you!" Eventually she and her husband emigrated to Mexico, where they worked as scientific translators until her death in 1966 (Bernstein 1992; Stull 1992).

Even the most innocent actions could mean trouble in the McCarthy era. In Chicago, Kay Way made the friendly gesture of letting a visitor from the Canadian Atomic Energy Program use her apartment while she was out of town. Unfortunately, this physicist was later convicted of treason in Canada. The fact that he had once used her apartment caused Way continuing security-clearance difficulties (Libby 1979).

Joan Hinton, who learned about the destruction of Hiroshima from a newspaper while on a bus to Santa Fe, strongly objected to military control of nuclear energy. She joined local political groups and became active in the movement to internationalize atomic energy. She wished to complete her graduate studies and selected the University of Illinois, since friends from the Omega group were going there, but Illinois either refused to admit women on principle or made it clear that a woman would not be welcome.

Fortunately, Enrico Fermi and Samuel K. Allison—a physics professor at the University of Chicago who was a member of the leadership at the Met Lab—invited Hinton to the Institute for Nuclear Studies at the University of Chicago, where she became part of the group of students and faculty surrounding Fermi. They met in the evenings to discuss new ideas in physics that happened to interest Fermi. In her second year she was awarded a fellowship and shared an office with Harold Agnew, future director of Los Alamos, at the very top of the lab. The attic next to them contained tanks of heavy water, which Allison needed badly for his experiments, but even he could not obtain a drop because the attic store belonged to the military.

Disgusted by what she perceived as the militarization of physics, Hinton decided to leave the field and travel to China, where her brother and his college roommate were working as dairy farmers for the Chinese Communists under Mao Tse-tung. Her brother enlisted the aid of Madame Sun Yat-sen to arrange his sister's travel. In 1948, Hinton left the United States for China on an adventurous trip that included being smuggled out of Beijing through Nationalist lines. Joan Hinton married her brother's roommate, Erwin Engst, and today lives in China, working as a designer of dairy farms (W. Hinton 1996; J. Hinton 1990).

Hinton still loves to work on any machine that needs repair, and she speaks casually of learning to maintain milk meters and making running repairs on her car. She does see other scientists from time to time. The Chinese-born American physicist and Nobel laureate C. N. Yang visited her in 1976 (Libby 1979). Her three children were raised in China speaking Chinese, but all of them now live in the United States. Hinton says that she is lucky to have witnessed the two most significant events of the twentieth century: the test of the first atomic bomb and the Maoist Revolution (J. Hinton 1990).

## AFTER THE PROJECT: CONTINUED CAREERS

With the end of the fighting, all the barriers to women that the war had lowered went back up. Many who wanted to continue doing scientific work in the newly created national laboratories, which grew out of the Manhattan Project sites, were told that they would have to take demotions (Butler et al. 1996). Others couldn't get jobs in the labs at all. Universities politely informed the female faculty members who had been teaching in place of men that their services would no longer be required. It seemed to some women that no matter how good a woman was, she was out of a job when the men came home.

Edith Quimby had made major contributions to radiobiology and had received many honors, but it was not until 1954, when she became president of a national scientific society, the American Radium Society

(Rossiter 1995), that she was finally promoted to full professor of radiology at Columbia University's College of Physicians and Surgeons. After her retirement in 1960, Quimby continued to lecture, write, and consult for almost twenty years. She also headed a scientific committee of the National Council on Radiation Protection and Measurements. She died in her New York City home in 1982 at the age of ninety-one (Rossi 1982).

Despite the obstacles, a number of female scientists managed to use the connections they had established during the war to further careers in the newly established national laboratories or in universities. Maria Mayer, for example, resumed her part-time appointment at Sarah Lawrence College for a year. When her husband moved to the University of Chicago as a full professor of chemistry, she joined the Enrico Fermi Institute, then the Institute for Basic Research, as a voluntary associate professor, performing all the duties of an associate professor for no salary, beyond a small income for being a consultant to the Metallurgical Laboratory. When the Met Lab was reorganized as Argonne National Laboratory, the director, Robert Sachs, who had been Mayer's student, offered her a half-time position with a real salary. In 1959, after publishing two books and developing the nuclear model that would later win her the Nobel Prize, Mayer became a full professor at the University of California at San Diego (Rempel 1993).

Chien-Shiung Wu retained her salaried position as a research associate at Columbia University and in 1952 became an associate professor of physics. Her later career at Columbia included fundamental experiments on nuclear beta decay that brought her many honors, including the Wolf Prize. In 1975 she became the first female president of the American Physical Society.

Kay Way began work at the National Bureau of Standards in 1947. There she continued to study nuclear properties and reactions, and she led the group that collected and systematized data on nuclear energy levels and reactions under the auspices of the National Academy of Sciences–National Research Council. This group produced *Nuclear Data Tables* and related publications, and Way served as editor for many years. She also played a leading role in establishing the consortium of universities that today conducts research at the facilities of Oak Ridge (Artna-Cohen et al. 1993).

Elda Anderson returned to her teaching position at Downer's College, but was then recruited by Oak Ridge National Laboratory, where she was placed in charge of health education and training. As a nuclear physicist, she played an important part in shaping the emerging discipline of health physics. The Elda E. Anderson Award, established in 1961, is presented to a young member of the Health Physics Society to recognize excellence in research or development, a discovery or invention, devotion to health physics, or significant contributions to the profession (Hickey 1995).

After only six months in Chicago, Isabella Karle returned to the University of Michigan as an instructor in chemistry in 1944. Work at the Met Lab was slowing as the production phase of the Manhattan Project accelerated, and Karle and her husband sought new challenges. Because of academic nepotism rules, however, they had difficulty finding a university at which they could both work. They accepted jobs with the Naval Research Laboratory in 1946, though they felt that this was a comedown from academic work.

Isabella Karle became head of the Naval Research Laboratory's X-ray Diffraction Section for the structure of matter. Her numerous honors include membership in the National Academy of Sciences, the Hildebrandt Award, the Garvan Medal of the American Chemical Society, and the presidency of the American Crystallographic Association. In 1994 she became the first woman to receive the Bower Award and Prize in Science. This $250,000 award from the Franklin Institute recognized her lifetime achievement in three-dimensional molecular modeling and determination of molecular structure, which influenced chemistry, biology, and medicine (Kathren et al. 1994; *American Men and Women of Science*; Roscher 1993; Silverman 1994).

Norma Gross accompanied her husband to New York City and was delighted when a friend from the Manhattan Project, who now headed the Chemistry Department at the brand-new Queens College campus of the CCNY system, offered her a job. She taught an occasional course and worked on research in organic chemistry until 1970, when the campus unionized and she was forced to teach full-time.

Leona Woods Marshall returned to Chicago as a research associate at the Institute for Nuclear Studies. She worked with Fermi on experiments that made the gradual transition from nuclear physics into the field known today as high-energy physics. In 1954 she was named an assistant professor, but after Fermi's death that year she moved to Los Alamos National Laboratory, then to Brookhaven National Laboratory, and then to New York University. In 1963 she went to the University of Colorado, where her research interests expanded to include astrophysics and cosmology. In '63 she also married Willard Libby, and she eventually joined him at UCLA, where she worked increasingly on problems of ancient climates and on writing books, one of which, *The Uranium People*, dealt with her experiences on the Manhattan Project (R. Howes 1993; Libby 1979).

Elizabeth "Diz" Graves continued to do experimental nuclear physics at Los Alamos after the war while raising a family. She and her family enjoyed life at Los Alamos, taking up skiing and bird-watching. She wrote poetry and sang well. Graves turned down a request from the White House to serve on the Atomic Energy Commission because her husband had recently been involved in a severe radiation accident and she did not feel that she could accept the new responsibilities (Divan 1991). Al Graves had been standing directly behind Louis Slotin on May 21, 1946, when Slotin accidentally assembled a critical mass of enriched uranium. Slotin threw himself onto the mass and separated the uranium hemispheres, absorbing a lethal dose of radiation but saving the lives of the other six men in the room. Graves developed some symptoms of radiation poisoning, but survived (Rapoport 1971).

Katherine Lathrop, after her stint at the Manhattan Project, became an associate biochemist at Argonne National Laboratory, where she was engaged in radium studies. Later she became a chemist at the University of Chicago. In 1954 she joined the staff at the Argonne Cancer Research Hospital at the University of Chicago as a research associate, and eventually became a professor, continuing work in nuclear medicine and allied areas. In the meantime she raised five children (OHRE 1996). "In November of 1952 my fifth child was born. (Incidentally, I am sometimes asked how I felt about working with radioactivity while pregnant. My answer is: I believed I was working in safe conditions. I have two controls [children born before their mother's participation in radiation research], three experimentals [children born while their mother was involved in radiation research], and ten grandchildren, all healthy and intelligent."

Miriam Posner Finkel continued her work on the biological effects of radiation as an associate biologist and then a senior biologist at Argonne National Laboratory, where her husband also worked. A determined woman, she stood up for her rights as a senior scientist in disputes with management over lab operations. She was widely recognized in her field and became a member of a subcommittee of the National Committee on Radiation Protection and Measurements, and a member of the Subcommittee on Internal Emitters of the National Academy of Sciences–National Research Council.

Hoylande Young became Director of Technical Information at Argonne National Laboratory, remaining in that position until her retirement in 1964. She was the first female division director at Argonne, and also the first chairwoman of the American Chemical Society. A fellow of the American Institute of Chemists and the American Association for the Advancement of Science, and a charter member of the American Nuclear Society, she received various honors, including the Chicago Section of the American Chemical Society Award for distinguished service. In 1963 she participated in the celebration of the twentieth anniversary of the first weighing of plutonium. In the same year, the Hoylande D. Young lecture series was established at Argonne in her honor, under the auspices of the Research Society of America. She married Crawford Failey, a friend since their days at the Toxicity Laboratory (Nicholls 1978; O'Neill 1979; *American Men and Women of Science*; Kathren et al. 1994).

When her work at Los Alamos ended, Agnes Stroud, one of the few Native American female scientists, went on to a distinguished career in radiation biology and cytogenetics. She moved from her home in a New Mexico pueblo to Argonne in 1946, received a Ph.D. in zoology from the University of Chicago in 1966, and continued working at Argonne until 1968. She worked as a microbiologist at several institutions before becoming director of the Department of Tissue Culture at the Pasadena Foundation for Medical Research, and as a senior biologist at the Jet Propulsion Laboratory in Pasadena. She then returned to Los Alamos as a radiobiologist in the Mammalian Biology Group at Los Alamos National Laboratory. Stroud was honored by the New York Academy of Sciences and the First Pan-American Cancer Cytology Congress (O'Neill 1979).

In October 1945, after working at the Met Lab and Hanford, Jane Hall became an associate physicist and assistant to the laboratory director (who was Enrico Fermi) at Argonne Laboratory of the Met Lab (Kathren et al. 1994). She didn't stay there long, however. Within two months she had returned to Los Alamos, where she worked for twenty-five years more. She pursued research in neutron physics, reactor development, X-ray crystallography, and cosmic radiation. Her technical work included measurement of absorption and activation in the Los Alamos plutonium-reactor spectrum (D. Hall et al. 1948). She served as an associate director of the laboratory, and as a consultant to and eventually a member of the Advisory Committee on Nuclear Materials Safeguards of the Atomic Energy Commission (Kathren et al. 1994; Sylves 1987). She was also named to the General Advisory Committee of the Atomic Energy Commission (Sylves 1987). She retired from Los Alamos in 1970 and died in Santa Fe in 1981.

Many other women extended their work on the Manhattan Project into careers—some along the lines of what they'd been doing for the project, some in new directions. Marge DeGooyer remained at Hanford, and she is still an expert at the art of setting up mass spectrographs. She is now retired, but does some consulting to industry. Frances Dunne, who worked with explosives at Los Alamos, went from there to the FBI. Marjorie Powell Shopp, who worked in the Los Alamos motor pool as a WAC, stayed on the Hill after the war, but as an electronics technician. Pat McAndrew, a sergeant in the WACs, was put in charge of the Los Alamos mail room, where she became renowned for correcting the grammar in outgoing letters that she was checking for security violations.

Ellen Weaver accompanied her husband, a graduate student in physics, to Cleveland in 1946. She supported both of them, first as a secretary, then as a technician. She found working as a technician particularly frustrating because people automatically assumed that she knew nothing. By the time her husband got his Ph.D. and moved to the Stanford Research Institute, she had an ulcer.

Over the next two years, while still working as a technician, she became fascinated by biology and managed to gain admission to the graduate program at Stanford. Her first formal biology course was a graduate course, but Weaver received a B. Her husband encouraged her to continue her studies, even though she had to work part-time. Her first course in genetics was scheduled when she could not attend, so her husband went instead and took careful notes. "We got an A!" she says. Weaver received her M.A. in genetics from Stanford in 1952 and her Ph.D. from Berkeley in 1959. She failed her orals the first time she took them, but with her husband's encouragement tried again, this time successfully (Weaver 1991).

After graduation, Weaver became a professor of biology at San Jose State University and conducted research on mechanisms of photosynthesis in algae, light-induced electron transport as monitored by means of electron paramagnetic resonance spectroscopy, evolution of photosynthesis, and remote sensing of chlorophyll in the oceans. She served as president of the Association for Women in Science.

Hilda Lefkowitz, a mathematician on the project in New York, left for Bell Labs in early 1945. Her first assignment at Bell Labs concerned the transport of magnetron tubes for radar to units at the front. The delicate and expensive vacuum tubes frequently failed after bouncing over rough roads, and Raymond Mindlin, the former Columbia professor whom Lefkowitz worked for, had been asked to design a container to protect them. His design consisted of a double box with eight helical springs supporting the inner box. Lefkowitz solved the differential equation Mindlin set up to determine the spring constants for maximum protection of the inner box. She prepared a numerical solution that later was used to check the operation of an early computer.

When the war ended, Mindlin returned to his position at Columbia, though he continued to consult for Bell Labs. Lefkowitz married, and became pregnant just as the labs moved to Murray Hill, New Jersey. She quit her job to care for the baby, which was followed by two more.

Mindlin, however, did not forget Lefkowitz. He called the young mother in Yonkers, New York, and persuaded her to take just one math course at Columbia. She continued to take roughly one course per semester. After all, she reasoned, lots of mothers left their children to play bridge; she was spending about the same amount of time taking math courses. The courses added up, and in 1948 Mindlin convinced the Columbia Mathematics Department that Lefkowitz's work on the vacuum-tube packing problem should count as a master's thesis. She later wrote a dissertation on crystals and received her Ph.D. in 1960. Lefkowitz's former professors at Hunter College tried to hire her, but she did not want to commute to Manhattan. They contacted Queens College, which eagerly hired her. She claims to have taught well and raised her children professionally, though at the expense of her publication rate. She is now retired after a successful career (H. Cooper 1996).

Florence Pachter Schulkin, a bookkeeper at Los Alamos, earned her bachelor's and master's degrees and became a speech therapist with the Los Angeles Unified School District (Schulkin 1991). Cora Cooper, who taught nursery school and kindergarten at Los Alamos, claims to have had twenty-two different careers, including licensed heavy-equipment operator, security officer, and writer of operating manuals (C. Cooper 1991).

As we have seen, marriages between project workers were common. Naomi Livesay and Eleanor Ewing, who operated computing machines at Los Alamos, married fellow Manhattan Project workers and followed career paths typical of many of their peers. Ewing (with the permission of her division leader) married Richard Ehrlich on July 7, 1945. The wedding was held in Fuller Lodge at Los Alamos. The bride wore a white eyelet dress and a veil made from Livesay's net petticoat. After a three-day honeymoon, the couple moved an extra cot into Ewing's room in the women's dormitory.

With the end of the war, the Ehrlichs finally got an apartment, though Richard's mother was not impressed when she visited—particularly since they had had two flat tires on the way to Albuquerque and had spent the night sheltering in a local jail. In February 1946, the couple moved to Cornell University, where Richard became a graduate student. Eleanor worked for one semester as an assistant in the Math Department and dropped out of graduate school. Shortly afterward, the first of the Ehrlichs' two children arrived, and she found herself too busy to return to a poorly paid job (Ehrlich 1991).

Naomi Livesay married Tony French in October of 1945. French, a experimental nuclear physicist with the British delegation, returned to England in February 1946, taking his wife with him. Klaus Fuchs, at that time head of the Theory Group at Harwell, knew Livesay from Los Alamos and offered her a job establishing a liaison between Harwell and Teddington, where a shared computer facility would be housed. Although the job would last for only two or three months, it was a civil-service position, so Livesay arranged an interview with the British Civil Service Commission. She discovered that the policy was to establish the grade of the job, determine the salary that would be offered to a man, and reduce it by 30 percent for a woman. She took the job anyway, but left when she and her husband moved to Cambridge, where her son and daughter were born.

Seven years later, the Frenches returned to the United States, where Tony joined the Physics Department at the University of South Carolina. Naomi taught for a year at Columbia College, but she had to hire a maid to care for her preschooler. By the time she paid the maid and the extra taxes required by the jump in tax brackets, only 16 percent of her salary remained. Men could deduct the cost of a maid as necessary for work, but a woman could not. After that year, she decided that working was just not worth the trouble and left mathematics (French 1990).

## In Conclusion

The Manhattan Project offered talented women the opportunity to pursue careers at the forefront of scientific research. Though the reason for hiring female scientists and technicians was that men were not available, and though there were no women among the leaders of the project, the bottom line is that women were welcomed into laboratories from which they would have been excluded in peacetime. They were encouraged to enter occupations and develop skills that had been considered distinctly unfeminine, and the encouragement extended to science. Women rose to the challenge.

Credit for research done by large groups generally goes to the leader of the group, at least in the popular press. Thus few women were mentioned in contemporary accounts of the development of the bomb. Lise Meitner, a female scientist who happened to be Jewish, merited an occasional sidebar, which is ironic, because Meitner refused to work on the Manhattan Project.

Subsequent accounts of the development of the bomb have also failed to give women the credit they deserve. Many of these books were written by women—Leona Marshall's book *The Uranium People* is an example—but they tend to be autobiographical, focusing on the author's work as a relatively junior scientist on the project and her interaction with the project leaders. (It's only fair to note that these autobiographies rarely emphasize the contributions of junior male scientists on the Manhattan Project unless the man worked very closely with the author or later became well known. Klaus Fuchs, for example, got much more ink in accounts written after he was convicted of espionage than he did in those written before.)

There have been a number of books and articles written about women at the Manhattan Project sites, but they emphasize the difficulties of domestic life and rarely stress the scientific work the women were doing. The women interviewed for these books and articles tended to talk about the scientific leaders with whom they worked rather than their own contributions. As for official project histories, they stress budget numbers and project milestones rather than the work of individual scientists.

For all these reasons, female scientists and technicians have all but disappeared from accounts of the Manhattan Project. It's clear, however, that they were active in nearly every aspect of the project's technical work. Given the opportunity, they demonstrated their ability to master complex technical issues and to make solid research contributions. Indeed, once they had tasted the excitement of research, many women decided to make science and technology their careers, and several became outstanding scientists.

Although many of the women who worked on the bomb later expressed concern about nuclear weapons and became involved with various movements aimed at controlling them, few of these women express guilt about their part in constructing the bomb. Their husbands, fiancés, and brothers were dying on the battlefields of Europe and Asia, and they felt it their duty to do anything possible to win the war. They took pride in the fact that they did it well.

Because the scientific community was the first to recognize the power and implications of nuclear weapons, many scientists, including a number of women, took the lead in educating the public, journalists, and politicians about their potential. Nearly all of them had been shocked by the massive destruction at Hiroshima and Nagasaki. Some quickly became active in groups working to prevent nuclear war, and some were caught in the disaster of the McCarthy era and lost jobs and reputations. Overall, there is no evidence that women's reactions to the atomic bomb and its implications differed markedly from those of their male colleagues.

The end of the war put pressure on women to leave scientific and technical fields. Although the January 1946 issue of *Mademoiselle* magazine listed electronics, medical research, science writing, and transatlantic aviation as new career areas for women (along with television acting), the reality was that society put pressure on women to return to the home. Female scientists and industrial workers were lauded along with GIs in the victory celebrations at the end of the war, but by 1950 the ideal woman was once again staying home with the children. In a looser labor market, women's salaries dropped rapidly in relation to those of their male peers. Not until the women's movement and the 1970s were women widely encouraged to become scientists and given the opportunity to practice their professions without fighting uphill battles.

Faced with economic and social disincentives to pursue their careers, many women dropped out of science. The surprising thing is that many did not. The opportunities presented to women by the Manhattan Project left the scientific community with a number of outstanding women who pursued successful careers as scientists and technicians. They are role models for the women who follow them today.

# Epilogue

OVER THE past fifty years, several generations of science students have enjoyed stories of technical derring-do from the Manhattan Project. Its success fueled government support for science and engineering, leading to the growth of research universities in the United States and a system of graduate education that is the envy of today's world. But it wasn't until the feminist movement of the 1970s that policy makers relearned the lesson of the Manhattan Project: that women represented a valuable but untapped resource of scientific talent.

The shock of the Soviet nuclear bomb and the launch of Sputnik had spurred federal efforts to encourage the study of science and mathematics; in the '70s, for the first time, such efforts were aimed specifically at young women and girls. In the life sciences, the effort succeeded admirably; in physics and engineering, however, women remain a small minority. Nonetheless, there have been a number of prominent female scientists who can serve as role models for girls. The rising generation grew up with Sally Ride and Diane Fossey, for example.

The story we have told here is important for two reasons. First, the contributions of women to the scientific and technological aspects of the Manhattan Project deserve recognition. Second, women aspiring to, or just considering, careers in physics and engineering should realize that their predecessors participated in the astonishing achievements of those disciplines during the development of the atomic bomb.

As we close this manuscript, we wonder, as we have throughout its writing, just how many female participants in the Manhattan Project we haven't found. Our detective work has sketched a picture of the women who worked on the project, but we know that there are others who have equally interesting stories and who made equally important contributions. We hope that they and their families will forgive us any omissions and let us know about their work. And we hope that other writers will take it upon themselves to complete the story we have begun.

# Appendix 1
## Female Scientific and Technical Workers in the Manhattan Project

| Name | Role in the Manhattan Project |
|---|---|
| Adams, Gayle | technician in Chemistry Department, University of California at Berkeley |
| Adolphson, Anna B. | laboratory technician, Health Division, Met Lab[1] |
| Allebach, Patricia L. | laboratory technician, Health Division, Met Lab |
| Amacker, Hope Sloan | WAC, Public Affairs Office, Hanford |
| Anderson, Elda E. "Andy" | biophysicist, Los Alamos; conducted fission measurements |
| Anslow, Gladys Amelia | chief, Communications and Information Section, Office of Field Service, Office of Scientific Research and Development |
| Argo, Mary Langs | theoretical physicist, Los Alamos; conducted fusion calculations |
| Arnette, Mary Ruth | scientific/technical worker, Clinton |
| Arnette, Thelma | scientific/technical worker, Clinton |
| Arson, Helen J. | laboratory technician, Health Division, Met Lab |
| Axelwright, Dorothy | scientific worker in autoradiography, Berkeley |
| Bachelder, Myrtle "Batch" | analytical chemist, Los Alamos; performed materials-purity determinations |
| Bacher, Jean | scientific assistant[2], Los Alamos; performed computations |
| Barbic, Anna | laboratory assistant (diener)[3], Chemistry Division, Met Lab |

NOTE: This list includes some women mentioned in the book who were involved primarily in nontechnical work.

| | |
|---|---|
| Barbic, Lorraine | technical worker, Chemistry Division, Met Lab |
| Bartlett, Helen Blair | ceramicist, MIT |
| Baumann, Marian S. | laboratory technician, Health Division, Met Lab |
| Baumbach, Nathalie Seifert | chemist (research assistant), Chemistry Division, Met Lab |
| Bell, Iris | WAC, security office, Hanford |
| Bench, Helen Landriani | chemist, Clinton |
| Bernstein, Elaine Katz | junior biologist, Health Division, Met Lab |
| Berry, Yvette | chemist and technician, Hanford; performed radioactive-sample preparation and analysis |
| Bethe, Rose | ran Housing Office, Los Alamos |
| Bisbeg, Virginia Hawley | technical worker, Met Lab |
| Blume, Amanda Pauline | *see Stein, Amanda Pauline B.* |
| Boggs, Elizabeth M. | explosives specialist, Explosives Research Lab |
| Bowman, Elinor A. | laboratory assistant, Chemistry Division, Met Lab |
| Boykin, Pearline | chemistry technician, Met Lab; one of the few black women in the Manhattan Project |
| Bradford, Rebecca | *see Divan, Rebecca Bradford* |
| Brode, Bernice | scientific assistant, Los Alamos; performed computations |
| Brown, Evelyn J. | laboratory assistant, Argonne Laboratory, Met Lab |
| Brown, Ruth Jones | scientific/technical worker, Clinton |
| Burke, Mary T. | junior physicist, Instruments Division, Met Lab |
| Burkhardt, Bernice Morgan | Document Division, Los Alamos |
| Burnett, Gladys S. | WAC, Tech Area, Los Alamos; switchboard operator |
| Butcher, Mitzi Mars | Employment Office, Hanford |
| Cahn, Ann | laboratory technician, Los Alamos |
| Caldwell, Eleanor F. | laboratory technician, Health Division, Met Lab |
| Calhoun, Opaline | technician, Chemistry Division, Met Lab |
| Calloway, Helena | laboratory assistant (diener), Chemistry Division, Met Lab |

| | |
|---|---|
| Campbell, Miriam White | technical draftsman, Los Alamos; worked on bomb assembly plans |
| Carney, Rose | research assistant, Met Lab |
| Carritt, Jeanne | technician, Medical Group, Los Alamos |
| Carson, Lillian | scientific/technical worker, Theoretical Group, Los Alamos |
| Carter, Dorothy E. | laboratory technician, Health Division, Met Lab |
| Casler, Ruth A. | laboratory technician, Chemistry Division, Met Lab |
| Chettkow, Sara B. | laboratory technician, Health Division, Met Lab |
| Chiotti, Wilma | ran classified library, Iowa State College |
| Clark, Joan R. | mathematics assistant, Carbide & Carbon Chemical Corporation |
| Coatie, Georgia Mason | worker, Hanford |
| Coleman, Virginia Spivey | chemist, Y-12 plant, Clinton |
| Cook, Mary Jane | scientific worker, Clinton; studied internal effects of radioactivity |
| Cooper, Cora F. | teacher, Los Alamos |
| Cortelyou, Ethaline | junior chemist, Information Division, Met Lab |
| Coveyou, Ida | scientific/technical worker, Clinton |
| Crawford, Lorraine | *see Heller, Lorraine Golden* |
| Crawford, Margaret | scientific/technical worker |
| Cunningham, Ida | laboratory assistant (diener), Chemistry Division, Met Lab |
| Curtis, Rozel E. | research associate, Met Lab |
| Dailey, Mary M. | junior chemist, Health Division, Met Lab |
| Daniels, Minnie M. | laboratory technician, Chemistry Division, Met Lab |
| Davies, Mrs. T. H. | physician, Clinton |
| Dayton, Jean Klein | electronics technician, Los Alamos; radiation detectors, weapons testing |
| DeGooyer, Marge Nordman | technician, Hanford |
| De Le Vin, Emma | scientific assistant, Los Alamos; performed computations |
| Divan, Rebecca Bradford | technician, Los Alamos; specialist in quartz-fiber balances |

| | |
|---|---|
| Dixon, Doris | laboratory assistant (diener), Chemistry Division, Met Lab |
| Duffield, Priscilla Greene | secretary to J. Robert Oppenheimer, Los Alamos |
| Dumas, Louise | WAC, Los Alamos |
| Dunne, Frances | technician and explosives supervisor, Los Alamos |
| Edwards, Roberta A. | laboratory technician, Health Division, Met Lab |
| Ehrlich, Eleanor Ewing | mathematician, Los Alamos |
| Elliott, Josephine | scientific assistant, Los Alamos; performed computations |
| Emmerson, Harryette H. | hospital dietitian, Los Alamos |
| Emmerson, Margaret M. | technician, Met Lab |
| Engst, Joan Hinton | *see Hinton, Joan* |
| Estabrook, Grace M. | statistician, Y-12 plant, Clinton |
| Evans, Marjorie Woodard | physical chemist, plutonium project, Berkeley |
| Ewing, Eleanor | *see Ehrlich, Eleanor Ewing* |
| Failey, Hoylande Young | senior chemist, Met Lab |
| Fair, Mary M. | laboratory technician, Health Division, Met Lab |
| Ferguson, Mrs. H. K. | CEO, H. K. Ferguson Company, company that built the thermal-diffusion plant at Clinton |
| Fermi, Laura | author; wife of Enrico Fermi, physicist and project leader |
| Fineman, Olga Giacchetti | *see Giacchetti, Olga* |
| Finkel, Miriam Posner | radiobiologist, Met Lab and Clinton |
| Fisher, Phyllis | author; wife of Los Alamos physicist |
| Flettinger, H. Anne | junior physicist, Physics/Metallurgy Division, Met Lab |
| Florin, Kathleen | *see Gavin, Kathleen "Kay"* |
| Ford, Mary Rose | health physicist, Clinton |
| Foreman, Beatrice | technician, Chemistry Division, Met Lab |
| Fornafelt, Elsie | analytical chemist, Iowa State College |
| Fortenberg, Rosellen B. | chemist and technician, Clinton; analytical chemistry |

| | |
|---|---|
| Foster, Margaret Dorothy | chemist and mineralogist, U.S. Geological Survey; conducted uranium and thorium analyses |
| Frankel, Mary | scientific assistant, Los Alamos; performed computations |
| Freeman, Elsie Mae | laboratory assistant, Chemistry Division, Met Lab |
| Freeman, Julia Elaine | laboratory assistant, Chemistry Division, Met Lab |
| French, Naomi Livesay | mathematician, Theoretical Division, Los Alamos |
| Frisch, Rose Epstein | biology technician, Los Alamos |
| Fuchs, Margaret | *see Melhase, Margaret Fuchs* |
| Gaevada, Betty | research assistant, Health Division, Met Lab |
| Gailor, Joanne Stern | clerk, K-25 plant, Clinton |
| Galuskin, Sylvia | *see Schwartz, Sylvia Galuskin* |
| Gaston, Evelyn O. | laboratory technician, Health Division, Met Lab |
| Gavin, Kathleen "Kay" | technician and clerk, Chemistry Division, Met Lab |
| Genzel, Hazel | scientific assistant, Los Alamos; performed computations |
| Giacchetti, Olga | technician, Chemistry Division, Met Lab |
| Gilbert, Miriam | physicist, Clinton |
| Gilbreath, Rachel | technician, Health Physics and Chemistry Divisions, Met Lab |
| Ginsburg, Mildred S. | *see Goldberger, Mildred G.* |
| Gish, Eleanor | scientific worker, Met Lab; conducted experimental pile calculations |
| Gnaedinger, Agnes | chemist, Met Lab |
| Gnjec, Katherine | laboratory assistant (diener), Chemistry Division, Met Lab |
| Goeppert-Mayer, Maria | *see Mayer, Maria Goeppert* |
| Goldberger, Mildred G. | research assistant and scientific assistant, Met Lab and Los Alamos |
| Golden, Lorraine | *see Heller, Lorraine Golden* |
| Goldfarb, Edith | junior physicist, Information Division, Met Lab |

| | |
|---|---|
| Goldowski, Nathalie Michel | metallurgist, Met Lab; studied corrosion in uranium cladding |
| Gordon, Pearl Leach | nurse, Los Alamos; worked with victims of radiation exposure |
| Graves, Elizabeth Riddle "Diz" | experimental physicist, Met Lab and Los Alamos; studied neutron scattering |
| Greenblatt, Sadelle T. | research assistant, Instruments Division, Met Lab |
| Greene, Priscilla | *see Duffield, Priscilla Greene* |
| Grieff, Lotti | chemist, Columbia University |
| Gross, Norma | chemist, Los Alamos |
| Guadagna, Lillian | technician, Met Lab |
| Hall, Helen | scientific/technical worker, Clinton |
| Hall, Jane Hamilton | physicist, Met Lab, Hanford, and Los Alamos |
| Hamilton, Jane | *see Hall, Jane Hamilton* |
| Hamilton, Kay | biologist, Argonne Laboratory, Met Lab |
| Hammer, Geneva Owen | coffee server in mess hall, Hanford |
| Hanig, Evelyn Kalichman | chemist, Clinton |
| Happer, Gladys Morgan | physician, Clinton |
| Harvey, Grace R. | laboratory technician, Health Division, Met Lab |
| Harvey, Kay | laboratory assistant in chemistry, Met Lab and Clinton |
| Hauk, Eleanor | *see Pomerance, Eleanor H.* |
| Heller, Lorraine Golden | research assistant, Chemistry Division, Met Lab |
| Heller, Minnie | junior biologist, Health Division, Met Lab |
| Herrick, Susan Chandler | chemist, Columbia University; studied uranium chemistry |
| Heydorn, Jane | WAC, Los Alamos |
| Hillard, Ruth | technician, Met Lab |
| Hillhouse, Dorothy | teacher, Los Alamos |
| Hinch, Josephine | research assistant, Met Lab |
| Hinton, Joan | experimental physicist, Los Alamos; worked on reactor design and construction |
| Hinton, Norma J. | laboratory technician, Chemistry Division, Met Lab |

| | |
|---|---|
| Hipple, Norma K. | laboratory technician, Health Division, Met Lab |
| Hitchcock, Kathleen | scientific/technical worker, Los Alamos |
| Holmberg, Reba | scientific/technical worker, Clinton |
| Holmes, Virginia S. | lab assistant, Berkeley |
| Hopkins, Joyce M. | research assistant, Health Division, Met Lab |
| Hornig, Lilli Schwenk | chemist, Los Alamos; studied plutonium chemistry and performed explosives research |
| Howe, Marilyn | research assistant, Met Lab |
| Hubbard, Cardea | laboratory assistant (diener), Chemistry Division, Met Lab |
| Hudson, Hazel | scientific assistant, Los Alamos; performed computations |
| Hughes, Helen R. | laboratory technician, Health Division, Met Lab |
| Hunter, Rosie M. | laboratory technician, Health Division, Met Lab |
| Hurwitz, Jean | *see Dayton, Jean Klein* |
| Inglis, Betty | scientific assistant and teacher, Los Alamos; performed computations |
| Jackson, Eugenia M. | laboratory technician, Health Division, Met Lab |
| Jackson, Jean A. | research assistant, Health Division, Met Lab |
| Janis, Frances T. | laboratory technician, Health Division, Met Lab |
| Jennings, Nellie | laboratory assistant, Chemistry Division, Met Lab |
| Jensen, E. Ann | clerk, Los Alamos |
| Jessee, Kathleen | laboratory technician, Health Division, Met Lab |
| Jette, Eleanor | author; wife of Los Alamos physicist |
| Johnson, Carrie | laboratory assistant (diener), Chemistry Division, Met Lab |
| Johnson, Elizabeth | scientific/technical worker, Clinton; worked on reactor operations |
| Johnson, Margaret L. | mathematician, Theoretical Division, Los Alamos |
| Johnson, Martha G. | laboratory technician, Health Division, Met Lab |

210 APPENDIX 1

| | |
|---|---|
| Johnson, Phyllis | scientific/technical worker, Clinton |
| Johnston, Melba | *see Robson, Melba Johnston* |
| Jones, Roberta Harvey | nurse, Los Alamos; worked with victims of radiation exposure |
| Jordon, Marilyn "Jodie" | laboratory assistant, Chemistry Division, Met Lab |
| Jupnik, Helen | physicist, Princeton University; measured resonant absorption of neutrons |
| Kalichman, Evelyn | *see Hanig, Evelyn Kalichman* |
| Karcher, Myrtle E. | laboratory technician, Health Division, Met Lab |
| Karle, Isabella Lugoski | associate chemist, Met Lab; studied transuranic chemistry |
| Katz, Elaine J. | *see Bernstein, Elaine Katz* |
| Katz, Sonia Weiner | secretary to the assistant director of the Chemistry Division, Met Lab |
| Keck, Margaret Ramsey | *see Ramsey, Margaret Keck "Peggy"* |
| Kenny, Ann | conducted calculations at Princeton University |
| Kingslow, Janice E. | laboratory technician, Health Division, Met Lab |
| Kirkley, Ada | *see Perry, Ada Kirkley* |
| Klicek, Olga M. | laboratory technician, Health Division, Met Lab |
| Koshland, Marian Elliott | junior chemist, Clinton |
| Koskosky, Adele | technician, Met Lab |
| Koziolek, Winifred T. | technician, Chemistry Division, Met Lab |
| Krinek, Helen E. | research assistant, Health Division, Met Lab |
| Kurath, Frances Wilson | mathematician, Los Alamos; conducted opacity calculations |
| Lane, Mary | secretary, Cyclotron Group, Los Alamos |
| Langer, Beatrice | scientific assistant, Los Alamos; performed computations |
| Larranaga, Gloria A. | laboratory technician, Health Division, Met Lab |
| Lathrop, Katherine A. | junior chemist, Health Division, Met Lab |
| Lawrence, Blanche J. | research assistant, Health Division, Met Lab |
| Lazarus, Marcella D. | traveling recruiter for Manhattan Project |

| | |
|---|---|
| Lear, Patricia | laboratory technician, Health Division, Met Lab |
| Lefkowitz, Hilda | mathematician, Columbia University |
| Lemke, Valda B. | laboratory helper, Chemistry Division, Met Lab |
| Lenoff, Gladys | nurse, Clinton; worked with victims of radiation exposure |
| Lewis, Beverley | chemist, Chemistry Division, Met Lab |
| Lewis, Eleanor | technician, Met Lab |
| Lewis, Fredrika | laboratory technician, Health Division, Met Lab |
| Lewitz, Elaine | see Novey, Elaine Lewitz |
| Leyshon, Emily | chemist, Y-12 plant, Clinton |
| Libby, Leona Woods Marshall | physicist, Met Lab and Hanford |
| Little, LaFern | laboratory technician, Health Division, Met Lab |
| Livesay, Naomi | see French, Naomi Livesay |
| Lomkee, Valda | laboratory helper, Met Lab |
| Lucas, Estelle B. | laboratory assistant (diener), Chemistry Division, Met Lab |
| Mabins, Lucy | laboratory assistant (diener), Chemistry Division, Met Lab |
| Maeder, Irma Dowd | WAC, Los Alamos; supply clerk in PX |
| Mandel, Marion | laboratory technician, Health Division, Met Lab |
| Manley, Kay | scientific assistant, Los Alamos; performed computations |
| Marcus, Elizabeth Painter | see Painter, Elizabeth Edith |
| Maris, Buena | supervisor of women's activities, Hanford |
| Marks, Edna K. | junior biologist, Health Division, Met Lab |
| Marshall, Laura A. | laboratory technician, Health Division, Met Lab |
| Marshall, Leona Woods | see Libby, Leona Woods Marshall |
| Martin, Alice | laboratory assistant (diener), Chemistry Division, Met Lab |
| Maxwell, Elizabeth | physician, Los Alamos |
| Mayer, Maria Goeppert | physicist, Columbia University and Los Alamos; conducted opacity calculations |

| | |
|---|---|
| McAndrew, Pat | WAC sergeant, Los Alamos; handled cash flow |
| McCamey, Creola Green | calutron operator, Clinton |
| McChesney, Marilyn | medical technician, Los Alamos |
| McDaniel, Mary Nell | scientific/technical worker, Clinton |
| McKenzie, LaVerna C. | laboratory technician, Health Division, Met Lab |
| McKibbin, Dorothy | head of Santa Fe office |
| McNunis, Ruth K. | laboratory technician, Health Division, Met Lab |
| Melhase, Margaret Fuchs | chemist, Berkeley |
| Meshke, Virginia | chemist and technician, Met Lab; specialized in plutonium chemistry |
| Meyers, Dorothy | secretary, Procurement Division, Los Alamos |
| Micese, Frances C. | laboratory technician, Health Division, Met Lab |
| Miller, Mary | technical draftsman, Met Lab |
| Miller, Mary L. | physical chemist, Los Alamos; in charge of a laboratory group |
| Minch, Josephine | research assistant in physics, Met Lab |
| Mokstad, Betty | technician, Chemistry Division, Met Lab |
| Monet, Marian Crenshall | chemical engineer, Clinton |
| Monk, Ardis | research assistant, Met Lab; performed computations |
| Mooney-Slater, Rose C. L. | physicist and crystallographer, Met Lab |
| Moore, Marietta C. | laboratory technician, Health Division, Met Lab |
| Morgan, Betty B. | research assistant, Health Division, Met Lab |
| Murray, Betty | technician, Solvent-Extraction Group, Met Lab |
| Nachtrieb, Mary | chemist, Los Alamos; specialized in plutonium chemistry |
| Neresin, Jean Parks | teacher, Los Alamos |
| Newman, Lorine G. | laboratory technician, Health Division, Met Lab |
| Newman, Mary Holiat | chemist, Clinton; worked on uranium-isotope separation |
| Nickson, Margaret Jane | physician, Biology Division, Met Lab |

| | |
|---|---|
| Noah, Frances E. | scientific assistant, Los Alamos; performed computations |
| Nordheim, Gertrude P. | theoretical physicist, Clinton and Los Alamos; worked on neutron diffusion |
| Norton, Sue G. | junior physicist, Health Division, Met Lab |
| Novey, Elaine Lewitz | laboratory assistant and technician, chemistry laboratory, Met Lab and Clinton |
| Nyden, Shirley | technician, Chemistry Division, Met Lab |
| O'Leary, Jean | administrative assistant to General Leslie Groves |
| Olsson, Virginia | managed Tennessee office of Manhattan Project |
| Painter, Elizabeth Edith | associate biologist, Health Division, Met Lab |
| Palevsky, Elaine Sammel | *see Sammel, Elaine* |
| Parrish, Mary Ellen | research assistant, Met Lab |
| Parsons, Aquila | technician, Met Lab |
| Patterson, Annie Lee | laboratory assistant (diener), Chemistry Division, Met Lab |
| Patterson, Pat | calutron operator, Clinton |
| Pellock, Helen | technician, Chemistry Division, Met Lab |
| Perley, Anne | biochemist, Los Alamos; monitored radiation exposure |
| Perlman, Ilsa | scientific/technical worker, Los Alamos |
| Perry, Ada Kirkley | chemical technician, Y-12 plant, Clinton |
| Perry, Sue | scientific/technical worker, Clinton |
| Peterson, Helen Pellock | *see Pellock, Helen* |
| Phillips, Anne | managed New York office of Manhattan Project |
| Phillips, Eleanor H. | laboratory assistant, Met Lab |
| Pierce, Dorothy | laboratory assistant (diener), Chemistry Division, Met Lab |
| Pierce, Elsie | WAC, Los Alamos; drove in motor pool |
| Pinckard, Marian | technician, Chemistry Division, Met Lab |
| Pomerance, Eleanor H. | technician and technical artist, Clinton and Berkeley; worked with calutrons |
| Pomeroy, Marian Smith | scientific/technical worker, Clinton |
| Porter, Lillie May | laboratory assistant, Chemistry and Health Divisions, Met Lab |

214   APPENDIX 1

| | |
|---|---|
| Porter, Victoria S. | laboratory technician, Health Division, Met Lab |
| Priest, Grace | chemist, Clinton |
| Prothrow, Annie | laboratory assistant (diener), Chemistry Division, Met Lab |
| Quimby, Edith Hinkley | biophysicist, Columbia University; studied medical effects of radiation |
| Ramsey, Margaret Keck "Peggy" | physicist, Los Alamos; worked on explosives |
| Ramsey, Mary | scientific/technical worker, Clinton |
| Rand, Margaret | research assistant, Health Division, Met Lab |
| Reace, Eleanor | chemist and metallurgist, MIT |
| Redmon, Jean | scientific/technical worker, Clinton |
| Rees, Mina | executive assistant, Applied Mathematics Board, Office of Scientific Research and Development |
| Reitz, Leola G. | laboratory technician, Health Division, Met Lab |
| Rhoades, Ruth P. | associate biologist, Health Division, Met Lab |
| Richardson, Elizabeth | scientific/technical worker, Clinton |
| Richmond, Marjorie L. | laboratory technician, Health Division, Met Lab |
| Riddle, Elizabeth | *see Graves, Elizabeth Riddle "Diz"* |
| Rizzoli, Ida | clerk, Accountable Property Department, Los Alamos |
| Roberg, Jane | physicist, Los Alamos; worked on fusion-weapon calculations |
| Roberts, June H. | research assistant, Instruments Division, Met Lab |
| Robinson, Donna | technician, Hanford |
| Robson, Melba Johnston | biomedical technician, Met Lab |
| Rona, Elizabeth | chemist, Rochester and Clinton; worked on polonium extraction |
| Rosen, Prisda K. | laboratory technician, Health Division, Met Lab |
| Rosenbluth, Arianna W. | mathematician, Los Alamos; |
| Rosenthal, Marcia White | biochemist, Met Lab |
| Rottman, Ruth M. | laboratory technician, Health Division, Met Lab |

| | |
|---|---|
| Rubinovich, Mrs. | technician, detector assembly, Met Lab |
| Russell, Leanne | biologist, Met Lab and Clinton |
| Sacher, Dorothea | nontechnical staff member, Met Lab |
| Sammel, Elaine | developer of optical instruments, Los Alamos |
| Sanderson, Margaret H. | laboratory technician, Health Division, Met Lab |
| Saunders, Addie | laboratory assistant, Chemistry Division, Met Lab |
| Savedoff, Lydia | physicist, Clinton |
| Schiedenhein, Arlene | commander of WACs assigned to the Manhattan Project |
| Schottmiller, Carolyn M. | laboratory technician, Health Division, Met Lab |
| Schulkin, Florence Pachter | WAC, Los Alamos; managed check-cashing |
| Schwartz, Bernice | mathematician, Columbia University |
| Schwartz, Sylvia Galuskin | WAC, Los Alamos; secretary |
| Sciofo, Madeleine A. | scientific/technical worker, Argonne Laboratory, Met Lab |
| Seifert, Nathalie | *see Baumbach, Nathalie Seifert* |
| Serber, Charlotte | technical-area librarian, Los Alamos |
| Shanklin, Mary M. | WAC, Los Alamos; telephone operator |
| Shaw, Margaret M. | supervisor of nurses, Hanford |
| Shazkin, Elky | physicist, Columbia University; was offered job on Manhattan Project, but could not accept because of nepotism rules |
| Shopp, Marjorie Powell | WAC, Los Alamos; drove in motor pool |
| Shor, Roberta | chemist, Clinton and Argonne Laboratory, Met Lab |
| Shupp, Selma | technician, Chemistry Division, Met Lab |
| Simonsen, Constance | technician, Los Alamos |
| Singleton, Acie | laboratory helper, Met Lab |
| Sites, Anna Mary | clerk, Accountable Property Department, Los Alamos |
| Skirmont, Ellen M. | laboratory technician, Health Division, Met Lab |
| Slater, Rose | *see Mooney-Slater, Rose C. L.* |
| Slaughter, Lulu L. | laboratory assistant, Met Lab |
| Smith, Alice Kimball | teacher and historian, Los Alamos |

| | |
|---|---|
| Smith, Barbara J. | technician, Clinton |
| Smith, Helene M. | laboratory technician, Health Division, Met Lab |
| Smith, Marie | cyclotron operator, Los Alamos |
| Sniegowski, Angeline | scientific assistant, Los Alamos; performed computations |
| Sohn, Madeline | research assistant, Chemistry Division, Met Lab |
| Speck, Lyda | physicist and technician, Los Alamos; worked on neutron spectroscopy |
| Stallings, Lenora | laboratory technician, Health Division, Met Lab |
| Stark, Helen | chemist, Clinton |
| Stark, Helene M. | laboratory technician, Health Division, Met Lab |
| Stark, Vivian | laboratory technician, Health Division, Met Lab |
| Steel, Gertrude F. | junior chemist, Health Division, Met Lab |
| Steele, Frances E. | WAC, Los Alamos; worked in finance and intelligence |
| Stein, Amanda Pauline B. | research assistant, Chemistry Division, Met Lab |
| Stewart, Leona | physicist, Los Alamos |
| Stone, Jerry | WAC, Los Alamos; worked in PX |
| Stroud-Schmink, F. Agnes Naranjo | microbiologist, Los Alamos |
| Stuart, Marie | chemist, Clinton |
| Summers, Mildred M. | chemical technician, Met Lab; one of the few black women in the Manhattan Project |
| Swift, Marguerite N. | junior physiologist, Health Division, Met Lab |
| Sypult, Edna May | laboratory helper, Met Lab |
| Tasseneau, Janet | biology technician, Met Lab |
| Taylor, Ethel | scientific assistant, Los Alamos; performed computations |
| Taylor, Moddie D. | associate chemist, Health Division, Met Lab |
| Teller, Augusta "Mici" | scientific assistant, Los Alamos; performed computations |
| Tenney, Edith | nurse, Los Alamos |

| | |
|---|---|
| Terzine, Camille M. | laboratory technician, Health Division, Met Lab |
| Thomson, Helen | technician, Chemistry Division, Met Lab |
| Thurman, Lorraine | laboratory helper, Met Lab, worked in chemical engineering |
| Tinsley, Ellen L. | teacher, Los Alamos |
| Tinsley, Mary P. | laboratory technician, Health Division, Met Lab |
| Tolmach, Emma | technician, Chemistry Division, Met Lab |
| Towle, Virginia | technician, Chemistry Division, Met Lab |
| Tracy, Kay (Mrs. A. S.) | secretary to Arthur Compton, director of Met Lab |
| Trujillo, Isabelle | clerk, Accountable Property Department, Los Alamos |
| Truslow, Edith C. | coauthor of official wartime history of Los Alamos |
| Tyree, Ella B. | laboratory technician and supervisor, Health Division, Met Lab |
| Uchiyamada, Hilda C. | research assistant, Met Lab and Los Alamos; performed computations |
| Wagner, Juanita "Billie" | analytical chemist, Clinton |
| Wagner, Roberta Shor | *see Shor, Roberta* |
| Walker, Evelyn S. | technician, Los Alamos; worked with metal oxides and plastics |
| Wallace, Dorothy | radiological physicist, Health Division, Met Lab |
| Wallace, Ida May | laboratory assistant (diener), Chemistry Division, Met Lab |
| Wallace, Jean F. | research assistant, Health Division, Met Lab |
| Walsh, Patricia D. | research assistant, Chemistry Division, Met Lab |
| Warner, Louise | research assistant, Health Division, Met Lab |
| Warshaw, Sylvia | research assistant, Chemistry Division, Met Lab |
| Watts, Ellen | technician, Chemistry Division, Met Lab |
| Way, Katharine "Kay" | physicist, Met Lab, Clinton, and Los Alamos; worked on reactor design |
| Weaver, Ellen C. | chemist, Clinton; performed fission fragment studies |

| | |
|---|---|
| Weiner, Sonia | *see Katz, Sonia Weiner* |
| White, Miriam | WAC sergeant, Los Alamos |
| Whiting, Mary J. | WAC, Los Alamos; drove in motor pool |
| Wilson, Elizabeth | technician, Los Alamos; worked in radioactive-materials handling |
| Wilson, Frances | *see Kurath, Frances Wilson* |
| Wilson, Jane | teacher and scientific/technical worker, Los Alamos |
| Winch, Josephine | research assistant, Met Lab |
| Wood, Nancy F. | junior physicist, Met Lab; designed and constructed radiation counters |
| Woods, Leona | *see Libby, Leona Woods Marshall* |
| Wooster, Marcia W. | laboratory technician, Health Division, Met Lab |
| Wright, Edith | scientific assistant, Los Alamos; performed computations |
| Wu, Chien-Shiung | nuclear physicist, Columbia University |
| Young, Gale | research associate, Theoretical Studies, Met Lab |
| Young, Hoylande Denune | *see Failey, Hoylande Young* |
| Zagaria, Susan | laboratory technician, Health Division, Met Lab |
| Zakoian, Aime | laboratory technician, Health Division, Met Lab |
| (last name unknown), Vera | mathematician, Los Alamos |

## Notes

1. Met Lab: University of Chicago Metallurgical Laboratory.

2. scientific assistant: These women were involved in performing numerical calculations in the theoretical division at Los Alamos. Many were the wives of scientists, and not all of them had scientific training or continued in science after the war.

3. laboratory assistant (diener): "Dieners" are referred to in Glenn Seaborg's journals; the word is German for "servant" or "maid." These positions were all classified as laboratory assistants (see e.g. p. 623), but they seem to have been—at most—only partly technical, to judge from a photo of "dieners" holding mops in *The Plutonium Story* (Kathren et al. 1994).

# Appendix 2
*Chronology*

1896      The French scientist Henri Becquerel discovers new penetrating radiations being emitted by some uranium compounds.

1898      In Paris, Marie Curie develops an instrument to measure the amount of ionization produced by this radiation. She shows that thorium as well as uranium emits these penetrating radiations, demonstrates that the radiation comes from individual atoms, and names the phenomenon "radioactivity." She names and classifies alpha, beta, and gamma "rays."

1901      Marie and Pierre Curie, together with Becquerel and Jean Perrin, put forward the idea that radioactivity is an atomic transformation process in which potential energy contained in the radioactive atom is liberated.

1905      The discovery of the mass–energy relationship of special relativity is published by Albert Einstein.

1906      The physicist Ernest Rutherford discovers the existence of the nucleus of the atom.

1913      The British historian and novelist H. G. Wells publishes *The World Set Free*, a science-fiction novel set in the future, which describes a new source of energy and weapons obtained from "atomic disintegration."

1929      In the United States, the Great Depression is initiated by the October stock-market crash.

1932      James Chadwick, an English physicist, announces the discovery of the neutron. In France, Irène and Frédéric Joliot-Curie independently make a similar discovery.

1934      Irène and Frédéric Joliot-Curie produce a radioactive isotope of a stable element, showing that radioactivity can be produced artificially.

         The Italian physicist Enrico Fermi produces new radioactivities by neutron bombardments of uranium and other materials, but does not recognize that they may be due to products of nuclear fission processes. He regards the new radioactivities only as evidence of the production of transuranic elements.

Ida Noddack, a German chemist, publishes a paper criticizing Fermi's work on transuranics and proposing that when heavy nuclei are bombarded with neutrons, the nuclei can break up into pieces that would be isotopes of known lighter elements. Her proposal is not taken seriously by Fermi and is ridiculed by Otto Hahn, an influential German chemist.

Leo Szilard, a Hungarian physicist, patents the concept of a nuclear chain reaction intermediated by neutrons.

1938 In December, Otto Hahn and Fritz Strassman demonstrate that the lower-mass element barium is in fact a product of the neutron bombardment of uranium.

In Sweden, Lise Meitner and Otto Frisch interpret Hahn and Strassman's results as the breakup of uranium nuclei after neutron bombardment. They describe the phenomenon and name it "nuclear fission," and calculate theoretically that a large release of energy occurs in fission.

1939

January

The Danish physicist Niels Bohr arrives in the United States to attend the Fifth Annual Conference of Theoretical Physics, and communicates the findings on the fission of uranium to American physicists.

Bohr predicts on theoretical grounds that only the isotope U-235 will fission with slow neutrons.

March

Szilard demonstrates that the irradiation of uranium disintegrates uranium nuclei with the emission of fast neutrons. Fermi, Szilard, and the Joliot-Curies, working independently, find that each fission produces two to three neutrons, implying that a nuclear chain reaction in uranium is theoretically possible.

August

After the physicists Leo Szilard, Eugene Wigner, and Edward Teller visit Albert Einstein, he signs a letter to President Roosevelt, informing him of German nuclear research and of the potential for a nuclear bomb.

September

German forces invaded Poland, marking the start of World War II.

October

President Roosevelt establishes an Advisory Committee on Uranium.

November
: The Advisory Committee on Uranium recommends that the U.S. government purchase graphite and uranium oxide for fission research.

December (approx)
: Irène and Frédéric Joliot-Curie demonstrate that the fission of uranium nuclei can lead to a nuclear chain reaction.

## 1940

March
: John Dunning and his colleagues at Columbia University demonstrate that fission is more readily produced in U-235, a rare isotope of uranium, than in the more plentiful U-238.

: Working in England, the Austrian scientist Otto Frisch and Rudolph Peierls, a German, calculate the critical mass required for a chain reaction with fast neutrons in pure uranium-235, and the result is communicated to the British and American governments.

May
: New elements with the atomic numbers 93 and 94 (later to be named neptunium and plutonium, respectively) are produced in the cyclotron at the University of California at Berkeley.

June
: The MAUD committee, in charge of Britain's uranium project, meets.

: Studies of uranium-isotope separation using uranium hexafluoride gas are conducted at Columbia University.

: The civilian National Defense Research Committee (NDRC) is established. The Advisory Committee on Uranium becomes a scientific subcommittee of the NDRC.

August
: The director of the Belgian company Union Miniere, which owns mines in the Belgian Congo that produce uranium, quietly directs the shipment of 200 steel drums containing 1,200 tons of high-grade uranium ore to the United States. The uranium will remain in a secluded storage facility on Staten Island, New York, for two years.

## 1941

February
: At Berkeley, the chemists Glenn Seaborg and Arthur Wahl conclusively establish the existence of element 94, which they later name plutonium.

March

In Berkeley, Seaborg's group demonstrates that plutonium is fissionable.

May

A National Academy of Sciences report addressing nuclear energy views the potential for the production of nuclear power favorably. The report does not address the potential for a nuclear weapon in any detail, but emphasizes the need for further research.

June

Vannevar Bush, head of the NDRC, becomes the director of the Office of Scientific Research and Development (OSRD), which is empowered to engage in large engineering projects as well as research. James Conant replaces Bush at the NDRC, which becomes an advisory body to the OSRD.

Germany invades the Soviet Union.

July

Emilio Segre and Glenn Seaborg measure the cross section of Pu-239 for fission by fast neutrons, and find that it is relatively large.

The British MAUD committee concludes that an atomic bomb is feasible.

In the United States, Bush and Conant receive the MAUD report.

October

Bush brings the report of the English MAUD committee to President Roosevelt. Roosevelt instructs Bush to find out whether a bomb can be built and how much it will cost. Bush receives permission to explore construction needs with the Army.

November

A National Academy of Sciences report agrees that constructing an atomic bomb is feasible.

December

President Roosevelt appoints a committee to undertake a uranium program under the S-1 Section of the OSRD. The first meeting is held, and the S-1 Executive Committee commits $400,000 to research on electromagnetic isotope separation.

Japanese planes bomb Pearl Harbor. The United States enters World War II. Germany and Italy declare war on the United States.

1942

January

Arthur H. Compton creates the Metallurgical Laboratory at the University of Chicago as a consolidated research center, with its emphasis on nuclear reactors and plutonium production and separation.

J. Robert Oppenheimer initiates a theoretical fast-neutron physics program at Berkeley.

March

The S-1 committee discusses priorities, and urges that the United States proceed simultaneously with all options for producing fissionable material.

April

Enrico Fermi relocates from Columbia University to the University of Chicago, where he starts building subcritical assemblies in preparation for constructing the first nuclear reactor.

Seaborg arrives in Chicago and starts work on developing an industrial-scale plutonium separation and purification program.

Percival Keith of the M. W. Kellogg Company begins the design of a gaseous-diffusion pilot plant for the separation of uranium isotopes.

May

The S-1 Executive Committee recommends that the project move to the pilot-plant stage and build one or two piles (nuclear reactors) to produce plutonium, and that electromagnetic, centrifuge, and gaseous-diffusion plants to produce uranium rich in U-235 be built.

June

The S-1 committee reports that an atomic bomb could be made at a cost of $100 million, perhaps by July 1944.

The OSRD recommends the establishment of an atomic-bomb production program.

President Roosevelt approves the S-1 Executive Committee's recommendation to proceed to the pilot-plant stage, and puts the Army in charge of plant construction and the development of a bomb. The OSRD continues to direct nuclear research, while the Army delegates the task of plant construction to the Corps of Engineers; an Army Corps of Engineers District is organized to take over and consolidate atomic-bomb development.

Production-pile designs are developed at the Met Lab in Chicago.

July

At Berkeley, Oppenheimer, Teller, and other physicists consider the design of an atomic bomb.

The first shipment of uranium (300 pounds) arrives at the Met Lab.

The Met Lab's Health Division is established to oversee environmental safety—in particular radiological health considerations—for the Manhattan Project.

August

The atomic-bomb program is established under the Manhattan District of the U.S. Army Corps of Engineers as the "Development of Substitute Materials" (DSM) project. The new organization is given the intentionally misleading name Manhattan Engineer District.

Seaborg's group produces a microscopic sample of pure plutonium by bombarding uranium in a cyclotron. Chemists develop efficient separation techniques for microgram quantities of plutonium—techniques regarded as suitable for industrial-scale applications.

Fermi's group at the Met Lab demonstrates an experimental pile with a projected k value—the net change in the number of thermal neutrons from one generation to the next—of close to 1.04; it's now certain that a nuclear chain reaction will be achieved.

American forces land on Guadalcanal in the Pacific.

September

A meeting of the S-1 Executive Committee discusses the need for a central fast-neutron laboratory, to be code-named Project Y.

The Military Policy Committee is established within the War Department under a civilian chairman to plan military policies relating to a nuclear-weapons program.

Colonel Leslie Groves is notified that he will take command of the weapons project code-named Manhattan: the Manhattan Engineer District, later to be known as the Manhattan Project. He is promoted to brigadier general six days later.

Groves buys the 1,200 tons of Belgian Congo uranium ore stored on Staten Island.

The S-1 Executive Committee visits Ernest Lawrence's Berkeley laboratory and recommends building an electromagnetic pilot plant and a section of a full-scale plant in Tennessee.

Groves selects Oak Ridge, Tennessee, as the site for a pilot plant and buys Site X, 52,000 acres of land on the Clinch River. This will become the site of the Clinton Engineer Works and subsequently the site of Oak Ridge National Laboratory.

The Manhattan Project is granted the highest emergency-procurement priority by the War Production Board.

Oppenheimer proposes the creation of a laboratory to study fast-neutron physics and develop designs for a bomb.

The U.S. Secretary of War, Henry Stimson, creates a Military Policy Committee to participate in decisions concerning the Manhattan Project.

October

General Groves visits the Met Lab and meets the leading scientists there; he orders that rapid engineering decisions be made on plutonium production.

General Groves puts E. I. Du Pont de Nemours in charge of the plutonium production project, and Du Pont agrees to build the chemical-separation plant at Oak Ridge.

General Groves decides to establish a separate scientific laboratory to design an atomic bomb, in addition to the factories for producing fissile materials. He asks Oppenheimer to head Project Y, the planned new central laboratory for weapons research and design.

Conant recommends dropping the centrifuge method of uranium isotope separation.

Compton recommends an intermediate pile at Argonne as part of the Met Lab.

November

General Groves and Oppenheimer visit the Los Alamos mesa in New Mexico and select it as the site for the bomb-development laboratory (Project Y). Oppenheimer is named the director of the Los Alamos lab.

Working twenty-four hours a day, Enrico Fermi's group at the Met Lab begins construction of the first nuclear reactor, Chicago Pile Number One (CP-1).

The decision is made not to build the plutonium production reactors at Clinton, as had been planned, but to find a more isolated area for safety reasons.

On the recommendation of Groves and Conant, the Military Policy Committee decides to skip the pilot-plant stage on the plutonium, electromagnetic, and gaseous-diffusion projects and go directly from the research stage to industrial-scale production.

The federal government buys 54,000 desolate acres in Sandoval County, New Mexico, for the modest price of $440,000. Two and a half years later, at the north end of the Jornada del Muerto in this area, the first nuclear explosion will take place.

The Allies invade North Africa.

December

After seventeen days of round-the-clock work, Fermi's group at the Met Lab in Chicago completes construction of the first nuclear reactor, CP-1. When the pile goes critical on December 2, Fermi and his colleagues achieve the first self-sustaining nuclear chain reaction.

Plutonium is produced in cyclotron operations and separated—a total of 500 micrograms.

President Roosevelt approves detailed plans for building production facilities and producing nuclear weapons.

1943

January

It is decided to build uranium and plutonium production plants.

General Groves demands that the first calutron at Clinton be finished by July 1.

Groves selects Hanford, Washington, as the site for the plutonium production facilities.

At Casablanca, President Roosevelt and Prime Minister Winston Churchill of Britain agree that they will demand the unconditional surrender of the Axis powers.

February

Construction begins at Clinton on buildings for Y-12, the electromagnetic uranium-isotope separation plant.

Groundbreaking for the X-10 plutonium pilot plant takes place at Clinton.

March

Los Alamos Laboratory is established in New Mexico to produce the fissionable components of bombs and to build and test suitable military nuclear weapons.

Oppenheimer moves to Los Alamos, researchers begin to arrive, and the lab starts operation.

April

At the beginning of the month, with the original building program for Los Alamos 96 percent complete, it is already apparent that the plans will be inadequate.

Los Alamos's initial organization and leadership plans are worked out.

A series of secret conferences concerning the capabilities of the proposed atomic bomb are held at Los Alamos for approximately 100 scientific staff members. Robert Serber, a physicist, gives lectures on the development of an atomic bomb to bring new staff up to speed; this leads to the subsequent publication of the "Los Alamos Primer": "The object of the project is to produce a practical military weapon in the form of a bomb in which the energy is released by a fast neutron chain reaction in one or more materials known to show nuclear fission."

Bomb design work begins at Los Alamos. Seth Neddermeyer, an experimental physicist recruited from the National Bureau of Standards, begins research on implosion.

May

Surveying begins for construction of the gaseous-diffusion uranium-isotope enrichment plant (K-25) at Clinton.

June

Site preparation for the K-25 plant begins at Clinton.

The Manhattan Engineer District moves its headquarters to Oak Ridge.

The first reactor-produced plutonium arrives at Los Alamos.

Captain William Parsons of the Navy arrives at Los Alamos as head of the Ordnance Division. He will direct research on the gun-assembly design for a nuclear weapon.

Working with cyclotron-produced plutonium, Emilio Segre's group discovers the occurrence of spontaneous fission in the isotope plutonium-240. The spontaneous fission rate is within the assembly-speed capability of a high-speed gun. This strongly suggests that a gun-design plutonium weapon will not work.

July

Oppenheimer reports that construction of a bomb may require three times as much fissionable material as had been thought nine months earlier.

The first explosive test of the implosion research program is conducted at Los Alamos.

August

Groundbreaking for the 100-B plutonium production pile at Hanford takes place.

Because of slow progress on gaseous diffusion and continuing uncertainty about the amount of U-235 needed for a bomb, General Groves decides to double the size of the Y-12 plant at Clinton. The construction staff at Clinton now exceeds 20,000 people.

The first Alpha electromagnetic uranium-separation unit begins operation at Clinton.

Construction begins on the cooling systems for the production reactors at Hanford. The construction staff at Hanford numbers about 5,000.

Roosevelt meets with Churchill and Prime Minister William King of Canada to outline cooperation in constructing an atomic bomb.

September

Construction begins on the K-25 facility at Clinton.

Oppenheimer suggests recruiting George Kistiakowsky, a Harvard chemist and the explosives research director at the OSRD, to contribute to an expanded effort to develop an implosion bomb.

Italy surrenders to Allied forces.

October

The first Alpha "racetrack" (calutron) for electromagnetic separation of uranium isotopes at Clinton is completed.

Project Alberta, the full-scale atomic-bomb delivery program, begins, with the physicist Norman Ramsey appointed to select and modify aircraft to deliver bombs.

November

Leading British experts on fission weapons travel to the United States to work on the Manhattan Project.

The X-10 pile at Clinton, the first true plutonium production reactor, goes critical and begins operation. X-10 will be able to produce plutonium in amounts that can be measured in grams; up to this time the world supply of plutonium has totaled about 2.5 milligrams, produced in cyclotrons.

The world's first sample of plutonium in metal form, still in tiny amounts, is produced at the Met Lab.

The U.S. Navy approves a plan to build a liquid thermal-diffusion pilot plant for isotopic enrichment of uranium.

The Manhattan Project governing board approves an accelerated implosion-research program intended to meet a six-month deadline for a usable bomb.

Berkeley scientists inform Los Alamos that U-235 releases its neutrons within a billionth of a second, which implies that a gun-type assembly device for a uranium bomb will work.

At Wright field in Ohio, a B-29 bomber undergoes the first modifications that will allow it to carry and deliver atomic bombs.

December

At Clinton, Y-12 is shut down after attempts to bring the first Alpha racetrack into operation fail.

At Los Alamos, Segre measures the spontaneous fission rate of U-235 and finds it lower than expected, suggesting that the planned gun assembly for the uranium bomb will be easier to realize.

Chemical separation of reactor-produced plutonium begins, using irradiated uranium from the X-10 reactor at Clinton.

## 1944

January

The second Alpha racetrack at Clinton is started, but has maintenance problems similar to those that disabled the first.

Continuing problems in developing suitable diffusion barriers lead General Groves to switch to a new type of barrier for production, causing months of delays in equipping K-25 at Clinton for operation.

Problems with the uranium gun-type bomb have been worked out at Los Alamos.

Construction begins at the Philadelphia Navy Yard on the Naval Research Laboratory's thermal-diffusion uranium-enrichment plant.

It is estimated that there will be enough plutonium for two bombs. (It had previously been predicted that barely enough uranium would be available to make one bomb.)

Groves and Oppenheimer decide to test a fission bomb, though they had not previously planned such a test. Groves specifies that the fissionable material must be recovered if the explosion fizzles, so the construction of Jumbo, a 214-ton steel container, is authorized.

February
: Y-12 sends 200 grams of U-235 to Los Alamos.

March
: At Clinton, the Beta building at Y-12 is completed.

: The first plutonium created by production plants is ready for use.

: Drop tests of dummy atomic bombs from modified B-29 bombers begin.

April
: Oppenheimer informs General Groves of the thermal-diffusion research in Philadelphia.

June
: General Groves contracts to have a liquid thermal-diffusion uranium-enrichment plant (S-50) built at Clinton within three months.

: Allied military forces invade Normandy in France.

July
: The decision is made to work on a calutron with a thirty-beam source for use in Y-12 at Clinton.

: Oppenheimer communicates the latest results of the Segre group's spontaneous-fission measurements to the Los Alamos scientific staff: The neutron emission rate for reactor-produced plutonium (which contained traces of the isotope Pu-240 in addition to fissile Pu-239) is too high for the gun-assembly method to work, as had been feared; the high neutron flux would cause predetonation. The plutonium gun bomb is abandoned and the implosion method for a plutonium bomb is made a top priority.

: Los Alamos reorganizes its administrative setup to direct the lab's resources toward the implosion method.

: Scientists at the Met Lab in Chicago issue the *Prospectus on Nucleonics*, concerning the international control of atomic energy.

August
: Vannevar Bush tells General George C. Marshall that small implosion-type plutonium bombs may be ready by mid-1945 and that a uranium bomb will almost certainly be ready by August 1945.

: Allied forces enter Paris. Allied intelligence learns that Germany has not developed an atomic bomb.

September

At Hanford, uranium is loaded into the first full-scale nuclear production reactor (the B pile). The pile goes critical and begins operation, but turns itself off during initial operation. Scientists, notably Chien-Shiung Wu, determine that the problem is caused by the presence of a radioactive fission product, Xe-135, which has an astonishingly large cross section for neutrons. Before plutonium production will be possible, the reactor must be modified to add extra reactivity to overcome the unexpected neutron loss.

At Clinton, K-25 is about half-built, but no satisfactory diffusion barriers have been produced. The Y-12 plant is operating, but at only about 0.05 percent efficiency.

The S-50 enrichment plant begins partial operation at Clinton, but leaks prevent substantial output. Total production of highly enriched uranium to this point has been only a few grams.

Roosevelt and Churchill sign an aide-mémoire pledging to continue bilateral research on nuclear technology.

At Wendover Field in Utah, the Air Force begins organizing the 509th Composite Group, which will deliver atomic bombs in combat. The 393rd Bombardment Squadron begins test drops with dummy bombs, called "pumpkins."

Vannevar Bush and James B. Conant advocate international agreements on nuclear research to prevent an arms race.

October

Oppenheimer approves plans for a bomb test at the Alamogordo bombing range. Groves subsequently approves.

November

At Clinton, Y-12's output has reached approximately forty grams of highly enriched uranium per day.

December

At Los Alamos, work begins on an implosion initiator for a solid-core bomb.

The first successful explosive lens tests are conducted at Los Alamos, establishing that an implosion bomb is feasible.

At Hanford, the D pile goes critical and large-scale production of plutonium starts.

At Hanford, the chemical-separation plants, called "Queen Marys," are finished. The processing of irradiated uranium slugs to separate plutonium begins.

At Hanford, the modified B pile is restarted.

The first bomb assembly (without explosive lenses or plutonium) is completed. The assembly design is named "Fat Man." Bomb assemblies are used for ground handling practice and airdrops.

## 1945

### January

At Clinton, uranium hexafluoride gas is introduced into the first stage of the K-25 plant, and the plant starts operation.

At Los Alamos, the "Dragon" experiment is conducted: A slug of U-235 hydride is dropped through a barely subcritical U-235 hydride assembly, creating the world's first prompt critical assembly.

### February

Los Alamos receives its first plutonium.

An incendiary air raid by Allied forces on Dresden, Germany, creates firestorms and kills 50,000.

U.S. Marines land on Iwo Jima.

A firebomb raid by U.S. forces burns a square mile of Tokyo.

Roosevelt, Churchill, and Premier Joseph Stalin of the Soviet Union meet at Yalta. The summit conference ratifies a divided postwar Europe.

### March

At Clinton, K-25 and other gaseous-diffusion plants are in operation.

At Clinton, the S-50 thermal-diffusion plant begins enriching uranium in significant amounts.

At Los Alamos, the first evidence of solid compression from implosion is observed.

Tokyo is firebombed by 334 B-29 bombers, killing at least 100,000 people.

### April

About twenty-five kilograms of U-235 and 6.5 kilograms of Pu-239 are now available. Producing enough reliable detonators remains a problem.

The committee responsible for selecting Japanese targets for atomic bombing meets for the first time.

Allied troops capture 1,100 tons of German uranium ore and several German nuclear scientists, who are sent to England.

President Roosevelt dies of a brain hemorrhage.

President Harry S. Truman, who had not known about the bomb, is briefed on the Manhattan Project.

American troops liberate the Nazi concentration camp at Buchenwald.

May

The German armed forces in Europe surrender. The United States and its allies celebrate V-E (Victory in Europe) Day.

American planes again firebomb Tokyo, causing 83,000 deaths.

June

Scientists at the Met Lab issue the Franck Report, which advocates international control of nuclear research and proposes a demonstration of the atomic bomb before its combat use.

A committee headed by James Byrnes, Secretary of State-designate, rejects the Franck Report's recommendation that the bomb be demonstrated before combat use. The committee recommends that the atomic bomb be dropped as soon as possible, that an urban area be the target, and that no warning be given.

The 509th Composite Group of B-29 bombers modified to deliver atomic bombs begins arriving on the Pacific island of Tinian.

July

The casting of the U-235 projectile for Little Boy, the first uranium bomb, is completed.

Little Boy bomb units, accompanied by the U-235 projectile, leave San Francisco for Tinian aboard the U.S.S. Indianapolis, a heavy cruiser.

At Los Alamos, the casting of explosive lenses for the Trinity test is completed.

Assembly of Gadget, the first nuclear explosive device to be tested at Trinity, begins.

Los Alamos scientists explode the world's first nuclear device (a plutonium implosion device) at the Trinity test site near Alamogordo, New Mexico, on July 16.

Hemispheres for the plutonium bomb, Fat Man, are fabricated.

General Groves drafts the directive authorizing the use of atomic bombs as soon as the weapons are available and the weather permits.

>  The 509th group begins flying practice missions over Japan. The group is ordered to attack Japan with an atomic bomb after about August 3.
>
>  The U.S.S. Indianapolis delivers the Little Boy bomb units to Tinian. The U-235 projectile arrives by air. The assembly of Little Boy is completed.
>
>  President Truman informs Stalin that the United States has tested a powerful new weapon.
>
>  Meeting in Berlin, representatives of the United States, Britain, and China issue the Potsdam Proclamation, calling for the prompt and unconditional surrender of Japan, but Japan refuses.

August

>  The United States drops a gun-model uranium bomb on the Japanese city of Hiroshima on August 6. This is the first detonation of a uranium bomb. President Truman announces the atomic bombing to the American public.
>
>  The Soviet Union declares war on Japan and invades Manchuria.
>
>  The United States drops an implosion-design plutonium bomb on the Japanese city of Nagasaki on August 9—the first detonation of an air-dropped plutonium bomb.
>
>  The Smyth Report, containing unclassified technical information on the bomb project, is released.
>
>  On August 10 or 11, General Groves reports that the second plutonium core will be ready for shipment on August 12 or 13, with a bombing possible on August 17 or 18, but President Truman orders a halt to atomic bombing until further orders are issued.
>
>  In the largest firebombing raid of the war, more than 1,000 B-29s drop 6,000 tons of conventional bombs on Japan.
>
>  Japan surrenders. World War II ends.

1946

March

>  Winston Churchill proclaims that an "iron curtain" has come down across Europe, after the Soviet Union obtains control or influence over the Eastern European countries it had liberated from the German army.

June

>  The Atomic Energy Act establishes the Atomic Energy Commission, with the aim of placing further development of nuclear technology under civilian control.

In Operation Crossroads at Bikini atoll in the Pacific, the United States conducts weapons effects tests on ships, using both airbursts and underwater bursts of plutonium bombs.

August

President Truman signs the Atomic Energy Act of 1946.

December

In accordance with the Atomic Energy Act of 1946, all atomic-energy activities are transferred, as of the end of 1946, from the Manhattan Engineer District to the newly created Atomic Energy Commission. The AEC replaces the Manhattan Project, bringing the project to a close. The Top Policy Group and the Military Policy Committee have already been disbanded. The Manhattan Engineer District is dissolved eight months later, followed at the end of 1947 by the NDRC and the OSRD, whose functions are transferred to the Department of Defense.

## Sources

Carter, Ashton B., John D. Steinbruner, and Charles A. Zraket, eds. (1987). *Managing Nuclear Operations*. Washington, D.C.: The Brookings Institution.

DOE Office of Environmental Management (October 1994). "Nuclear Age Timeline," poster supplement and resource guide. Washington, D.C.: U.S. Department of Energy. Also at http://www.em.doe.gov/timeline.

Hellmans, Alexander and Bryan Bunch (1988). *The Timetables of Science*. New York: Simon & Schuster.

Gosling, F. G. (August 1990). *The Manhattan Project: Science in the Second World War*, DOE/MA-01417P, Energy History Series. Washington, D.C.: The U.S. Department of Energy.

Miller, Richard L. (1986). *Under the Cloud: The Decades of Nuclear Testing*. New York: The Free Press division of Macmillan, Inc.

Rhodes, Richard (1986). *The Making of the Atomic Bomb*. New York: Simon and Schuster.

# References

Adams, Frank and Forrest L. Weaver (1944). *The Home Front.* Santa Monica, CA: Weaver Publishing Company.

Adamson, June N. (December 12, 1994). Letter to C. Herzenberg.

*American Men and Women of Science 1992–1993, 18th Edition.* New Providence, NJ: R. R. Bowker. Also earlier editions.

*American Men of Science* (earlier edition of *American Men and Women of Science*). New Providence, NJ: R. R. Bowker.

Argo, Harold (July 18, 1991). Telephone interview with R. Howes.

Aspray, William (1990). *John von Neumann and the Origins of Modern Computing.* Cambridge, MA: MIT Press.

Bachelder, Myrtle C. (November 1991). Questionnaire and telephone interview with R. Howes.

Badash, Lawrence (1979). *Radioactivity in America: Growth and Decay of a Science.* Baltimore: The Johns Hopkins University Press.

Bailey, Martha J. (1994). *American Women in Science: A Biographical Dictionary.* Denver: ABC-CLIO.

Bainbridge, Kenneth T. (1974). "Orchestrating the Test" in *All in Our Time*, edited by Jane Wilson. Chicago: Bulletin of the Atomic Scientists.

Barnett, Shirley B. (1988). "Operation Los Alamos" in *Standing By and Making Do: Women of Wartime Los Alamos*, edited by Jane S. Wilson and Charlotte Serber. Los Alamos, NM: Los Alamos Historical Society.

Barr, E. Scott (1964). "The Incredible Marie Curie and Her Family," *Physics Teacher* 2, pp. 251–259.

Barschall, Heinz (October 30, 1991). Telephone interview with R. Howes.

Bederson, Benjamin (1996). Conversation (January 20) and telephone interview (February 12) with R. Howes.

Bell, Iris (October 11, 1991). Letter.

Bennis, Warren and Patricia Ward Biederman (1997). *Organizing Genius: The Secrets of Creative Collaboration.* New York: Addison Wesley.

Bernal, J. D. (1954). *Science in History.* London: Watts & Co.

Bernstein, Melvin (September 1992). Telephone interview with R. Howes.

Bigland, Eileen (1957). *Madame Curie.* New York: Criterion Books.

Bond, Peter D. and Chellis Chasman (July 1998). "Gertrude Scharff Goldhaber." *Physics Today* 51, No. 7, pp. 82–83.

Bonner, Francis T. (1993). "Melba Newell Phillips (1907–)" in *Women in Chemistry and Physics: A Biobibliographic Sourcebook*, edited by Louise S. Grinstein, Rose K. Rose, and Miriam H. Rafailovich. Westport, CT: Greenwood Press.

Borman, Stu (July 17, 1995). "Chemists Reminisce on 50th Anniversary of the Atomic Bomb." *Chemical and Engineering News*, pp. 53–63.

Bradbury, Norris (1980). "Los Alamos—The First 25 Years" in *Reminiscences of Los Alamos 1943–1945*, edited by Lawrence Badash, Joseph O. Hirschfelder, and Herbert P. Broida. Vol. 5 of *Studies in the History of Modern Science*, edited by Robert S. Cohen, Erwin N. Hiebert, and Everett I. Mendelsohn. Boston: D. Reidel.

Brinkley, David (1988). *Washington Goes to War*. New York: Alfred A. Knopf.

Brode, Bernice (1980). "Tales of Los Alamos" in *Reminiscences of Los Alamos 1943–1945*, edited by Lawrence Badash, Joseph O. Hirschfelder, and Herbert P. Broida. Vol. 5 *of Studies in the History of Modern Science*, edited by Robert S. Cohen, Erwin N. Hiebert, and Everett I. Mendelsohn. Boston: D. Reidel.

Brown, Anthony Cave and Charles B. MacDonald (1977). *The Secret History of the Atomic Bomb*. New York: The Dial Press/James Wade.

Brush, Stephen G. (1985). "Women in Physical Science: From Drudges to Discoverers." *The Physics Teacher* 23, pp. 11–19.

Burkhardt, Lewis C. (October 2, 1991). Questionnaire.

Burnett, Gladys S. (November 1991). Questionnaire.

Butler, Margaret, Caroline L. Herzenberg, and Jane Andrew (1996). *Women in Scientific and Technical Positions at Argonne National Laboratory—Women and Argonne, Partners in Science: Five Decades of Memories and Meaning: 1946–1996*. Argonne, IL: Argonne National Laboratory (published in conjunction with the Second Technical Women's Symposium).

Campbell, Miriam White (November 6, 1991). Telephone interview with R. Howes.

Carritt, Jeanne Brooks (January 15, 1992). Letter and questionnaire.

Catlett, J. Stephen, ed. (1987). *A New Guide to the Collections in the Library of the American Philosophical Society*. Philadelphia: American Philosophical Society.

*CHEMTECH* (January 1989). "Reich Exile Emerges as Heroine," p. 17.

Clark, Ronald W. (1961). *The Birth of the Bomb*. London: Phoenix House.

Compton, Arthur Holly (1956). *Atomic Quest*. New York: Oxford University Press.

Cooper, Cora F. (November 1991). Questionnaire.

Cooper, Hilda Lefkowitz (June 3, 1996). Telephone interview with R. Howes.

Coveyou, Ida M. (November 1991). Questionnaire.

Crane, P. W. and W. O. Switzer (September 5, 1945). "Technical Department - Procurement and Training of Non-Exempt Personnel." Memorandum for file from the E. I. Du Pont De Nemours & Company archives at the Hagley Museum and Library in Wilmington, DE.

Crawford, Deborah (1969). *Lise Meitner, Atomic Pioneer*. New York: Crown.

Creutz, Ed (November 5, 1991). Telephone interview with R. Howes.

Curie, Eve (1937). *Madame Curie* (translated by Vincent Sheean). New York: Doubleday, Doran & Company.

CWP. Web site on Contributions of Twentieth-Century Women to Physics (http://www.physics.ucla.edu/~cwp).

Dampier, William Cecil (1949). *A History of Science and Its Relations with Philosophy and Religion.* Cambridge, U.K.: Cambridge University Press.

Dash, Joan (1973). *A Life of One's Own: Three Gifted Women and the Men They Married.* New York: Harper and Row.

Davis, J. L. (1995). "The Research School of Marie Curie in the Paris Faculty, 1907–14." *Annals of Science* 52, pp. 321–355.

Dayton, Jean Klein Hurwitz (November 1991). Questionnaire.

Debus, Allen G., ed. (1968). *World Who's Who in Science.* Chicago: Marquis Who's Who.

DeGooyer, Marge (October 1995). Telephone interview with R. Howes and script of talk on her work.

Divan, Rebecca Bradford (October 1991). Telephone interview with R. Howes.

DOE Roadmap (1996). From OHRE Web site (http://www.ohre.doe.gov).

Dumas, Louise (1991). Questionnaire.

Dunne, Frances (December 1991). Questionnaire and telephone interview with R. Howes.

Ehret, Charles (March 26, 1992). Private communication to C. Herzenberg.

Ehrlich, Eleanor Ewing (July 18, 1991). Telephone interview with R. Howes. Also subsequent written communications, including draft of memoirs.

Einstein, Albert and Mileva Maric (1992). *The Love Letters,* edited and with an introduction by Jurgen Renn and Robert Schulmann, translated by Shawn Smith. Princeton, NJ: Princeton University Press.

Emmerson, Harryette Hunter (November 1991). Questionnaire.

Estabrook, Grace McCammon (November 1991). Questionnaire, supplemental material, telephone interview with R. Howes, and letter.

Fermi, Laura (1954). *Atoms in the Family.* Chicago: The University of Chicago Press.

Fermi, Laura (1980). "The Fermis' Path to Los Alamos" in *Reminiscences of Los Alamos 1943–1945,* edited by Lawrence Badash, Joseph O. Hirschfelder, and Herbert P. Broida. Boston: D. Reidel.

Fineman, Olga Giacchetti (April 29, 1996). "A Look Back in Time," presented at the Argonne National Laboratory Second Women's Technical Symposium.

Finkel, Miriam. Transcript of interview (September 7, 1976) from OHRE Web site (http://www.ohre.doe.gov). Telephone interview with R. Howes (October 5, 1995).

Fisher, Phyllis K. (1985). *The Los Alamos Experience.* Tokyo: Japan Publications.

Fitch, Val L. (1974). *All In Our Time: The Reminiscences of Twelve Nuclear Pioneers,* edited by Jane Wilson. Chicago: The Bulletin of the Atomic Scientists.

Fleck, George (1993). "Gladys Amelia Anslow (1892–1969)" in *Women in Chemistry and Physics: A Biobibliographic Sourcebook,* edited by Louise S. Grinstein, Rose K. Rose, and Miriam H. Rafailovich. Westport, CT: Greenwood Press.

Folkart, Bruce. A. (November 13, 1986). "Leona Marshall Libby Dies; Sole Woman to Work on Fermi's 1st Nuclear Reactor." *Los Angeles Times.*

Ford, Mary Rose (April 17, 1979). Transcript of interview on OHRE Web site (http://www.ohre.doe.gov).

Fortenberg, Rosellen Bergman (November 1991) Questionnaire.

Freedman, Mel (February 1, 1991). Private communication. Also subsequent telephone conversations with C. Herzenberg and R. Howes.

French, Naomi Livesay (March 1990). Interview. Also subsequent correspondence with R. Howes.

Friedell, Hymer Louis (September 28, 1994). Interviewed on behalf of the Department of Energy Office of Human Radiation Experiments. On OHRE Web site (http://www.ohre.doe.gov).

Friedlander, Gerhart (September 9, 1997). Telephone interview with R. Howes.

Fries, Sylvia Doughty (1992). *NASA Engineers and the Age of Apollo*. Washington, D.C.: NASA document number SP-4104.

Frisch, O. R. (1970). "Lise Meitner, 1878–1968." *Biographical Memoirs of Fellows of the Royal Society* 16, pp. 405–420.

Gailar, Joanne Stern (October 27, 1990). Letter to C. Herzenberg.

Gerber, Michele Stenehjem (1992). *On the Home Front: The Cold War Legacy of the Hanford Nuclear Site*. Lincoln, NE: University of Nebraska Press.

Gerber, Michele Stenehjem (1993). *The Hanford Site: An Anthology of Early Histories*. Richland, WA: Westinghouse Hanford Company.

Gerber, Michele Stenehjem (February 1994). Private communication with C. Herzenberg.

Giroud, Francoise (1986). *Marie Curie, a Life*, translated by Lydia Davis. New York: Holmes and Meier.

Goldowski, N. (January 9, 1945). "Effect of Cl Concentration on the Corrosion of Beryllium." Oak Ridge National Laboratory classified document No. 45-1-148, as recorded in ORNL request for destruction of classified documents. On OHRE Web page (http://www.ohre.doe.gov).

Golonka, M. Cieslak, J. Roziewicz, J. Starosta, and K. G. Takhadze (1994). "Jadwiga Szmidt (1889–1940), a Pioneer Woman in Nuclear and Electrotechnical Sciences." *American Journal of Physics* 62, pp. 947–948.

Goodman, Mark (February 1996). "Cold War Human Radiation Experiments: A Legacy of Distrust." *APS News*.

Goodwin, Irwin (September 1991). "Happer Walks into the Caldron of DOE's Top Research Position." *Physics Today*, pp. 65–66.

Goodwin, Irwin (1991). Private communication with C. Herzenberg.

Gordon, Pearl Leach (November 7, 1991). Questionnaire and telephone interview with R. Howes.

Gosling, F. G. (August 1990). *The Manhattan Project: Science in the Second World War*. Washington, D.C.: U.S. Department of Energy. Energy History Series, DOE/MA-01417P.

Graetzer, Hans, and D. L. Anderson, eds. (1971). *The Discovery of Nuclear Fission: A Documentary History*. New York: Van Nostrand–Reinhold.

Graf, William L. (1994). *Plutonium and the Rio Grande*. New York: Oxford University Press.

Granados, Mary (May 8, 1997). Private communication to C. Herzenberg.

Greenbaum, Leonard (1971). *A Special Interest: The Atomic Energy Commission, Argonne National Laboratory, and the Midwestern Universities.* Ann Arbor: The University of Michigan Press.

Gross, Norma (September 23, 1997). Telephone interview with R. Howes.

Groueff, Stephane (1967). *Manhattan Project: The Untold Story of the Making of the Atomic Bomb.* Boston: Little, Brown.

Groves, Leslie R. (1962). *Now It Can Be Told.* New York: Harper & Brothers.

Gurer, Denise W. (1995). "Pioneering Women in Computer Science." *Communications of the ACM* 38, pp. 45–54.

Guttman, Les (March 13, 1991). Private communication.

Hacker, Barton C. (1987). *The Dragon's Tail: Radiation Safety in the Manhattan Project, 1942–1946.* Berkeley, CA: University of California Press.

Hales, Peter Bacon (1997). *Atomic Spaces: Living on the Manhattan Project.* Urbana: University of Illinois Press.

Hall, David B., Jane H. Hall, and A. H. Spano (October 1948). "Effective Absorption and Activation Cross Section in the Los Alamos Plutonium Reactor Spectrum." Presented in the technical program of the AEC Information Meeting, Argonne National Laboratory.

Hall, J. H. (1944). "Hazards in event of product being dispersed." Hanford report, document 7-635.

Hampel, Clifford A., ed. (1968). *The Encyclopedia of the Chemical Elements.* New York: Reinhold.

Hanford Works (1945). "Hazards of Waste Storage at Hanford Works." Report No. HW-14058. On OHRE Web site (http://www.ohre.doe.gov).

Hanford Works Design Engineering Department (January 1, 1948). Personnel listing. On OHRE Web site (http://www.ohre.doe.gov).

Hanson, David J. (February 13, 1995). "Energy Department Shuffle Probably Will Expel Some Research." *Chemical and Engineering News*, p. 27.

Happer, William (October 24, 1991). Letter to C. Herzenberg.

Harvey, Roger A., M.D. (October 23, 1945). Letter to Dr. Margaret Nickson. On OHRE Web site (http://www.ohre.doe.gov).

Hawkins, David (1961). *Manhattan District History, Project Y, The Los Alamos Project: Vol. 1. Inception Until August 1946.* Los Alamos, NM: Los Alamos Scientific Laboratory.

Hellmans, Alexander and Bryan Bunch (1988). *The Timetables of Science.* New York: Simon and Schuster.

Henriksen, Paul W. (June 25, 1986). Interview with Lilli Hornig, OH-128. Cited in Hoddeson et al. 1993.

Henriksen, Paul W. (1994). Private communication.

Herrick, Susan Chandler (November 4, 1991). Questionnaire and telephone interview with R. Howes.

Herzenberg, Caroline L. and Ruth H. Howes (1993). "Women of the Manhattan Project." *Technology Review* 96, pp. 32–42.

Hickey, Eva Eckhert (1995). "1995 Elda E. Anderson Award." *Health Physics* 69, pp. 875–876.
Hinton, Joan (December 1990). Telephone interview, letters, and conversation with R. Howes.
Hinton, William (August 29, 1990). Letter.
Hoddeson, Lillian, Paul W. Henriksen, Roger A. Meade, and Catherine Westfall (1993). *Critical Assembly: A Technical History of Los Alamos during the Oppenheimer Years 1943–1945.* New York: Cambridge University Press.
Hoffman, Joseph G. and Louis H. Hempelmann (1957). "Estimation of Whole-Body Radiation Doses in Accidental Fission Bursts." *The American Journal of Roentgenology, Radium Therapy, and Nuclear Medicine* 77.
Hogsett, Vic (1980). "Pat McAndrew: Being Nosey Was My Job." *Atom* 17, p. 11.
Holl, Jack M., Richard G. Hewlett, and Ruth R. Harris (1997). *The History of Argonne National Laboratory.* Urbana, IL: University of Illinois Press.
Holton, Gerald (1994). "Of Love, Physics and Other Passions: The Letters of Albert and Mileva." *Physics Today* 47, August pp. 23–29 and September pp. 37–43.
Hornig, L. (1993) Private communication with C. Herzenberg.
Horning, Beth (1993). "Trying Times at Oak Ridge and Beyond." *Technology Review* 96, pp. 38–39.
Howes, Jane Reed. Numerous conversations and personal papers.
Howes, Ruth H. (1993). "Chien-Shiung Wu (1912–)" in *Women in Chemistry and Physics: A Biobibliographic Sourcebook,* edited by Louise S. Grinstein, Rose K. Rose, and Miriam H. Rafailovich. Westport, CT: Greenwood Press.
Howes, Ruth H. (1993). "Leona Woods Marshall Libby (1919–1986)" in *Women in Chemistry and Physics: A Biobibliographic Sourcebook,* edited by Louise S. Grinstein, Rose K. Rose, and Miriam H. Rafailovich. Westport, CT: Greenwood Press.
Howes, Ruth H. and Caroline L. Herzenberg (1993). "Women in Weapons Development: The Manhattan Project" in *Women and the Use of Military Force,* edited by Ruth H. Howes and Michael R. Stevenson. Boulder, CO: Lynne Rienner Publishers.
*Industrial Research* (October 1974). "Scientist of the Year," pp. 29–30.
Jensen, E. Ann (1991). Questionnaire.
Jette, Eleanor (1977). *Inside Box 1663.* Los Alamos, NM: Los Alamos Historical Society.
Johnson, Margaret L. (November 1991). Questionnaire and telephone interview with R. Howes.
Jones, L. M. (1990). "Intellectual Contributions of Women to Physics" in *Women of Science: Righting the Record,* edited by G. Kass-Simon, Patricia Farnes, and Deborah Nash. Bloomington: Indiana University Press.
Jones, Roberta Harvey (November 1991). Questionnaire.
Jones, V. C. (1985). *United States Army in World War II—Special Studies—Manhattan: The Army and the Atomic Bomb.* Washington, D.C.: Center of Military History, U.S. Army.

Julian, Maureen M. (1986). "Isabella L. Karle and a New Mathematical Breakthrough in Crystallography." *Journal of Chemical Education* 63, pp. 66–67.

Julian, Maureen M. (1990). "Women in Crystallography" in *Women of Science: Righting the Record,* edited by G. Kass-Simon, Patricia Farnes, and Deborah Nash. Bloomington: Indiana University Press.

Jungk, Robert (1960). *Brighter Than a Thousand Suns.* New York: Penguin Books.

Karle, Isabella Lugoski (November 7, 1991). Telephone interview with R. Howes.

Kass-Simon, G., Patricia Farnes, and Deborah Nash, eds. (1990). *Women of Science: Righting the Record.* Bloomington: Indiana University Press.

Kathren, Ronald L., Jerry B. Gough, and Gary T. Benefiel, eds. (1994). *The Plutonium Story: The Journals of Professor Glenn T. Seaborg, 1939–1946.* Columbus, OH: Battelle Press.

Katz, Sonia (1999). Private communication with C. Herzenberg.

Kauffman, George B. and Jean-Pierre Adloff (1993). "Marguerite Catherine Perey (1909–1975)" in *Women in Chemistry and Physics: A Biobibliographic Sourcebook,* edited by Louise S. Grinstein, Rose K. Rose, and Miriam H. Rafailovich. Westport, CT: Greenwood Press.

Keck, Margaret Ramsey (February 21, 1991). Letter to R. Howes.

Kevles, Daniel J. (1978). *The Physicists.* New York: Alfred A. Knopf.

Kisieleski, Walt. (October 25, 1993, December 8, 1994, and September 27, 1996) Private communications to C. Herzenberg.

Krstic, Dord (1991). "Mileva Einstein-Marić" in *Hans Albert Einstein: Reminiscences of His Life and Our Life Together* by Elizabeth Roboz Einstein. Iowa City: Iowa Institute of Hydraulic Research, The University of Iowa.

Kurath, Dieter (May 31, 1990). Private communication to C. Herzenberg.

Lamont, Lansing (1965). *Day of Trinity.* New York: Atheneum.

Lane, Mary (November 1991). Questionnaire and telephone interview with R. Howes.

Langer, Beatrice (March 5, 1996). Telephone interview with R. Howes.

Langevin, M. (1964). "Radioactivity and Nuclear Physics, 1930 to 1940" in *History of Science: Science in the Twentieth Century,* edited by René Taton, translated by A. J. Pomerans. New York: Basic Books.

Lathrop, Katherine (January 26, 1995). Taped interview on OHRE Web site (http://www.ohre.doe.gov).

Lazarus, Marcella D. (December 26, 1993). Letter.

Leng, Herta R. (1991). "Pioneer Woman in Nuclear Science." *American Journal of Physics* 59, p. 584.

Lenoff, Gladys (November 1991). Questionnaire.

Levine, Arnold, S. (1987). *The Apollo Program: Science/Engineering Personnel Demand Created by a Federal Research Mission.* Washington, D.C.: U.S. Congress, Office of Technology Assessment.

Libby, Leona Marshall (1979). *The Uranium People.* New York: Crane Russak and Charles Scribner's Sons.

Lisco, Hermann, Miriam P. Finkel, and Austin M. Brues (1947). "Carcinogenic Properties of Radioactive Fission Products and of Plutonium." *Radiology* 49, No. 3, p. 361. Also on OHRE Web site (http://www.ohre.doe.gov).
Loeb, Paul (1986). *Nuclear Culture: Living and Working in the World's Largest Atomic Complex.* Philadelphia: New Society Publishers.
Loriot, Noelle (1991). *Irène Joliot-Curie.* Paris: Presses de la Renaissance.
*Los Alamos Science* (1995). "Polonium Human-Injection Experiments." No. 23, p. 196.
Los Alamos Scientific Laboratory. *Los Alamos 1943–1945; The Beginning of an Era,* LASL-79-78. Los Alamos, NM: Los Alamos Scientific Laboratory.
Lubkin, Gloria (1971). "Chien-Shiung Wu, The First Lady of Physics Research." *Smithsonian* 1, pp. 52–56.
*Mademoiselle* magazine (January 1946). "Mlle. Merit Awards," p. 132.
Maeder, Elmo M. (1991). Questionnaire.
Malone, Willie (October 30, 1995). Private communication with C. Herzenberg.
Manley, Kathleen Baird (March 1996). Telephone interview with R. Howes.
Manley, Kathleen E. B. (April 1990). "Women of Los Alamos During World War II: Some of their Views." *New Mexico Historical Review,* pp. 251-266.
Marshak, Ruth (1988). "Secret City" in *Standing By and Making Do: Women of Wartime Los Alamos,* edited by Jane S. Wilson and Charlotte Serber. Los Alamos, NM: Los Alamos Historical Society.
Marshall, C. L. (August 13, 1947). Office cover memo for "Effect of Single Dose X-Ray to The Nail Fold Area of Human Subjects" by Margaret Nickson. On the OHRE Web site (http://www.ohre.doe.gov).
Martin, Murray J., Norwood B. Gove, Ruth M. Gove, and Agda Artna-Cohen (1993). "Katharine Way (1903–)" in *Women in Chemistry and Physics: A Bio-bibliographic Sourcebook,* edited by Louise S. Grinstein, Rose K. Rose, and Miriam H. Rafailovich. Westport, CT: Greenwood Press.
McCamey, Creola Green (November 19, 1996). Telephone interview with R. Howes.
McGrayne, Sharon Bertsch (1993). *Nobel Prize Women in Science.* New York: Birch Lane Press.
McKay, Alwyn (1984). *The Making of the Atomic Age.* Oxford, England: Oxford University Press.
McKibbin, Dorothy (1980). "I'll Take Manhattan." *Atom* 17, pp. 7–9.
McKibbin, Dorothy (1988). "109 East Palace" in *Standing By and Making Do: Women of Wartime Los Alamos,* edited by Jane S. Wilson and Charlotte Serber. Los Alamos, NM: Los Alamos Historical Society.
McMurray, Emily J., Jane Kelly Kosek, and Roger M. Valade III, eds. (1995). *Notable Twentieth-Century Scientists.* New York: Gale Research/ITP.
McNulty, William (May 10, 1946). "World's Most Famed Scientists, En Route to Los Alamos Project, Go Through Ancient City Office." *Santa Fe New Mexican.*
Meitner, Lise and O. R. Frisch (1939). "Disintegration of Uranium by Neutrons: a New Type of Nuclear Reaction." *Nature* 143, p. 239.

Metropolis, Nicholas and E. C. Nelson (1982). "Early Computing at Los Alamos." *Annals of the History of Computing* 4, pp. 348-357.
*Monsanto* magazine (February 1946).
Morgan, Karl Ziegler (1991). Telephone interviews.
Moulton, Grace (July 1982). "Rose C. L. Slater." *Physics Today* 35, p. 72.
Mrozowski, Stanislaw (March 1991). Conversations with R. Howes.
Myers, Dorothy (November 1991). Questionnaire and supplemental material.
Newman, Mary Holiat. (February 22, 1991). Telephone interview and subsequent correspondence.
Nicholls, Jewel (February 1978). "Profile—Hoylande D. Young Failey." *Women Chemists Committee Newsletter.* Washington, DC: American Chemical Society.
Nichols, K. D. (1987). *The Road to Trinity.* New York: William Morrow.
Nickson, J. J. (1951). "Blood Changes in Human Beings Following Total-Body Irradiation" in *Industrial Medicine on the Plutonium Project,* edited by Robert S. Stone. Vol. 4, No. 20, of National Nuclear Energy Series. New York: McGraw-Hill.
Nier, Alfred (1989). "Some Reminiscences of Mass Spectrometry and the Manhattan Project." *Journal of Chemical Education* 66, pp. 385–388.
Noddack, Ida Tacke (1934). "Über das Element 93." *Zeitschrift Angewandte Chemie* 47, pp. 653–655.
Nordheim, Lothar W. (1971). "Old Times and New Horizons." *Oak Ridge National Laboratory Annual Report 1970–71.* Oak Ridge, TN: Oak Ridge National Laboratory. Reprinted in *Oak Ridge National Laboratory Review,* fall 1976.
Noyes, W. A., ed. (1948). *Chemistry: A History of the Chemistry Components of the National Defense Research Committee, 1940–1946.* Boston: Little, Brown.
Oak Ridge National Laboratory (1992). *Oak Ridge National Laboratory Review* 25, Nos. 3 and 4 (combined).
O'Brien, Henrietta (January 28, 1993). Letter to C. Herzenberg.
OHRE (Office of Human Radiation Experiments archives). Web site prepared by the Department of Energy (http://www.ohre.doe.gov).
O'Neill, Lois Decker (1979). *The Women's Book of World Records and Achievements.* New York: Da Capo Press.
Opfell, Olga (1986). "Pale Glimmer of Radium: Marie Curie" in *The Lady Laureates: Women Who Have Won the Nobel Prize,* 2nd Ed., pp. 147–164. Metuchen, NJ: Scarecrow Press.
Opfell, Olga (1986). "Triumph and Rebuff: Irène Joliot-Curie" in *The Lady Laureates: Women Who Have Won the Nobel Prize,* 2nd. Ed., pp. 165–182. Metuchen, NJ: Scarecrow Press.
Parker, H. M. (1945). Various memorandums in the Du Pont archives at the Hagley Museum and Library in Wilmington, DE.
Parker, Herbert (June 1979). "Interview with H. M. Parker." On OHRE Web site (http://www.ohre.doe.gov).
Parkinson, Claire L. (1985). *Breakthroughs: A Chronology of Great Achievements in Science and Mathematics 1200–1930.* London: Mansell Publishing Ltd.

Peierls, Rudolph (1989). "Reflections on the Discovery of Fission." *Nature* 342, pp. 852–854.
Perley, Anne (November 1991). Questionnaire.
Perry, Alfred M. (November 1991). Questionnaire.
Peshkin, Murray (1990). Private communication with C. Herzenberg.
Phillips, Melba (March 1990). Conversations.
Pierce, Elsie (January 29, 1991). Telephone interview with R. Howes.
Pomerance, Herbert (1991). Letter (March 3) and telephone interview with R. Howes (November 20).
Powers, Faye. Clippings from personal files.
Prosser, C. Ladd, with contributions by E. E. Painter, Hermann Lisco, Austin M. Brues, Leon O. Jacobson, and M. N. Swift (1947). "The Clinical Sequence of Physiological Effects of Ionizing Radiation in Animals." *Radiology* 49, No. 3, p. 299.
Pycior, Helena M. (1987). "Marie Curie's 'Anti-natural Path': Time Only for Science and Family" in *Uneasy Careers and Intimate Lives: Women in Science, 1789–1979*. New Brunswick, NJ: Rutgers University Press.
Quimby, Edith H. (August 1984). Oral History for the Advisory Committee on Human Radiation Experiments. Vignettes of early workers in radiation (transcripts of a videotape series). U.S. Department of Health and Human Services, Public Health Service, Food and Drug Administration. On OHRE Web site (http://www.ohre.doe.gov).
Quinn, Susan (1995). *Marie Curie: A Life*. New York: Simon and Schuster.
Rapoport, Roger (1971). *The Great American Bomb Machine*. New York: Ballantine Books.
Rayner-Canham, Marelene and Geoff Rayner-Canham (March 5, 1994). Letter to C. Herzenberg.
Rayner-Canham, M. F. and G. W. Rayner-Canham (1990). "Pioneer Women in Nuclear Science." *American Journal of Physics* 58, pp. 1036–1043.
Rayner-Canham, M. F. and G. W. Rayner-Canham (1992). *Harriet Brooks: Pioneer Nuclear Scientist*. Montreal: McGill–Queen's University Press.
Reace, Eleanor H. (January 13, 1994). Letter.
Reid, Robert (1974). *Marie Curie*. New York: New American Library.
Rempel, Trudy D. (1993). "Maria Gertrude Goeppert Mayer (1906–1972)" in *Women in Chemistry and Physics: A Biobibliographic Sourcebook*, edited by Louise S. Grinstein, Rose K. Rose, and Miriam H. Rafailovich. Westport, CT: Greenwood Press.
"Reports of the Various Laboratories to the X-10 Projects Advisory Committee" (August 21, 1945). Summaries and transcripts from meeting held in A. H. Compton's office at the University of Chicago. On OHRE Web site (http://www.ohre.doe.gov).
Rhodes, Richard (1986). *The Making of the Atomic Bomb*. New York: Simon and Schuster.
Rhodes, Richard (1992). Introduction in *The Los Alamos Primer* by Robert Serber. Berkeley, CA: University of California Press.

Richards, H. T. and Lyda Speck (1947). "Range Distribution of $U^{235}$ Fission Fragments in Photographic Emulsions." *Physical Review* 71, p. 141.

Richards, H. T., Lyda Speck, and I. H. Perlman (1946). "Neutron Spectra." *Physical Review* 70, p. 118.

Rife, Pat (1980). "Lise Meitner (1878–1968). Part 1: The Early Years." *Association for Women in Mathematics Newsletter* 10, No. 3, pp. 8–13.

Rife, Pat (1980). "Lise Meitner (1878–1968). Part 2: The Mathematical Interpretation of Nuclear Fission." *Association for Women in Mathematics Newsletter* 10, No. 4, pp. 9–14.

Rife, Patricia (1995). *Lise Meitner and the Dawn of the Nuclear Age.* Boston: Birkhäuser.

Ringo, Roy (February 21, 1991). Private communication.

Rizzoli, Ida J. (December 31, 1991). Letter.

Robson, Andrew (1995). Private communications to C. Herzenberg and R. Howes.

Robson, David (1995). Private communications to C. Herzenberg.

Robson, Judith (1995). Private communications to C. Herzenberg.

Rodden, Clement J. (1989). "Before NBL." *Journal of the Institute of Nuclear Materials Management.* New Brunswick Laboratory fortieth-anniversary issue, Vol. 18, No. 1.

Roensch, Eleanor Stone (1993). *Life Within Limits.* Los Alamos, NM: Los Alamos Historical Society.

Rona, Elizabeth (1978). *How It Came About: Radioactivity, Nuclear Physics, Atomic Energy.* Oak Ridge, TN: Oak Ridge Associated Universities.

Roscher, Nina Matheny (1993). "Isabella Helen Lugoske Karle (1921–)" in *Women in Chemistry and Physics: A Biobibliographic Sourcebook,* edited by Louise S. Grinstein, Rose K. Rose, and Miriam H. Rafailovich. Westport, CT: Greenwood Press.

Rossi, Harald H. (December 1982). "Edith Hinkley Quimby," *Physics Today* 35, pp. 71–72.

Rossiter, Margaret W. (1982). *Women Scientists in America: Struggles and Strategies to 1940.* Baltimore: The Johns Hopkins Press.

Rossiter, Margaret W. (1995). *Women Scientists in America: Before Affirmative Action: 1940–1972.* Baltimore: The Johns Hopkins Press.

Sacher, Dorothea (December 19, 1995). Private communication to C. Herzenberg. "Early Met Lab Days," recorded recollections by George Svihla, Mildred Summers, H. Kubischek, W. Kisieleski, W. P. Norris, and F. Wasserman, May 11, 1965.

Sacher, George A. Jr. (1952). "A Sentimental History of Site B." *Argonne News* 4, No. 5, pp. 4–5.

Sachs, Robert G. (1979). "Maria Goeppert Mayer" in *Biographical Memoirs of the National Academy of Sciences* 50, pp. 310–328.

Sanger, S.L. (1989). *Hanford and the Bomb: An Oral History of World War II.* Seattle: Living History Press.

Schriesheim, Alan (1992). Private communication to C. Herzenberg.
Schwartz, George and Philip W. Bishop (1958). *Moments of Discovery Vol. 2: The Development of Modern Science*. New York: Basic Books.
Schwartz, Samuel, Elaine J. Katz, Lillie Mae Porter, Leon O. Jacobson, and Cecil James Natson (July 10, 1946). "Studies of the Hemolytic Effect of Radiation." Metallurgical Laboratory report CH-3760.
Schwartz, Sylvia Galuskin (November 1991). Questionnaire.
Seaborg, Glenn T. (July 10, 1970). "Reminiscences on the Development of Some Medically Useful Radionuclides." Remarks at the Seventeenth Annual Meeting of the Society of Nuclear Medicine, Washington, D.C. On the OHRE Web site (http://www.ohre.doe.gov).
Seaborg, Glenn T. (February 1977). *History of Met Lab Section C-I: April 1942 to April 1943*. Lawrence Berkeley Laboratory publication PUB 112, University of California at Berkeley.
Seaborg, Glenn T. (May 1978). *History of Met Lab Section C-I: May 1943 to April 1944*. Lawrence Berkeley Laboratory publication PUB 112 Vol. 2, University of California at Berkeley.
Seaborg, Glenn T. (May 1979). *History of Met Lab Section C-I: May 1944 to April 1945*. Lawrence Berkeley Laboratory publication PUB 112, Vol. 3, University of California at Berkeley.
Seaborg, Glenn T. (1989). "Silver, Copper, and 'Honest-to-God Copper.'" *Journal of the Institute of Nuclear Materials Management* 18, No. 1, pp. 16–17.
Seaborg, Glenn T. (June 6, 1996). Private communication to C. Herzenberg.
Seidel, Bob (1992). "The Italian Navigator Has Landed in the New World: Secret Race Won with Chicago's Chain Reaction." On the Web (http://lib-www.lasl.gov/pubs/lasl50th/12-11-92.html).
Seidel, Robert W. (1993). "Evolving from calculators to computers." On the Web (http://lib-www.lanl.gov/pubs/lanl50th/9-24-93.html).
Serber, Charlotte (1988). "Labor Pains" in *Standing By and Making Do: Women of Wartime Los Alamos*, edited by Jane S. Wilson and Charlotte Serber. Los Alamos, NM: Los Alamos Historical Society.
Serber, Robert (1996). "Theoretical Studies at Berkeley" in *Behind Tall Fences*. Los Alamos, NM: Los Alamos Historical Society.
Shanklin, Mary M. (1991). Questionnaire.
Shopp, Marjorie Powell (November 1991). Questionnaire.
Sietmann, Richard (1988). "False Attribution: A Female Physicist's Fate." *Physics Bulletin* 39, No. 8, pp. 316–317.
Silverman, Edward R. (January 10, 1994). "Naval Lab 'Experimentalist' Honored with Bower Award." *The Scientist*, p. 4.
Sime, Ruth Lewin (1989). "Lise Meitner and the Discovery of Fission." *Journal of Chemical Education* 66, pp. 373–375.
Sime, Ruth Lewin (1994). "Lise Meitner in Sweden 1938–1960: Exile from Physics." *American Journal of Physics* 62, pp. 695–701.
Sime, Ruth Lewin (1996). *Lise Meitner: A Life in Physics*. Berkeley: University of California Press.

Smith, Alice Kimball (1965). *A Peril and a Hope: The Scientists' Movement in America 1945–47*. Chicago: The University of Chicago Press.
Smith, Alice Kimball (1988). "Law and Order" in *Standing By and Making Do: Women of Wartime Los Alamos*, edited by Jane S. Wilson and Charlotte Serber. Los Alamos, NM: Los Alamos Historical Society.
Smith, Alice Kimball and Charles Weiner, eds. (1980). *Robert Oppenheimer: Letters and Recollections*. Cambridge, MA: Harvard University Press.
*Smith Alumnae Quarterly* (November 1945). "Smith Scientists Help Win the War," pp. 3–5.
Smith, Barbara S. (November 1991). Questionnaire.
Smyser, Dick (1992). *Oak Ridge 1942–1992: A Commemorative Portrait*. Oak Ridge, TN: Oak Ridge Community Foundation.
Smyth, H. D. (1977). *The Smyth Report*, reprinted in *The Secret History of the Atomic Bomb*, edited by Anthony Cave Brown and Charles B. MacDonald. New York: The Dial Press.
Snyder, Thomas D. (1993). *120 Years of American Education: A Statistical Portrait*. Washington, D.C.: National Center for Education Statistics.
Sopka, Katherine R. (1984). "Women Physicists in Past Generations" in *Making Contributions: An Historical Overview of Women's Role in Physics*, edited by Barbara Lotze. College Park, MD: American Association of Physics Teachers.
Sorenson, Don (1995). Interviews conducted at Hanford and provided to the authors.
Speck, Lyda (October 5, 1991). Letter, questionnaire, and supplemental material.
Spindel, William (November 30, 1993). Letter and telephone interview.
Spradley, Joseph L. (1989). "Women and the Elements: The Role of Women in Element and Fission Discoveries." *The Physics Teacher* 27, pp. 656–662.
Stachel, John et al. (1987). *The Collected Papers of Albert Einstein: The Early Years: 1879–1902*. Princeton, NJ: Princeton University Press.
Starke, K. (1979). *Journal of Chemical Education* 56, p. 771.
Steele, Frances E. (1991). Questionnaire.
Stull, John (September 30, 1992). Telephone interview with R. Howes.
Svec, Harry (1995). Telephone interview with R. Howes (October 10) and subsequent letter.
Svoronos, Paris (1993). "Irène Joliot-Curie (1897–1956)" in *Women in Chemistry and Physics: A Biobibliographic Sourcebook*, edited by Louise S. Grinstein, Rose K. Rose, and Miriam H. Rafailovich. Westport, CT: Greenwood Press.
Svoronos, Soraya (1993). "Marie Sklodowska Curie (1867–1934)" in *Women in Chemistry and Physics: A Biobibliographic Sourcebook*, edited by Louise S. Grinstein, Rose K. Rose, and Miriam H. Rafailovich. Westport, CT: Greenwood Press.
Sylves, Richard T. (1987). *The Nuclear Oracles: A Political History of the General Advisory Committee of the Atomic Energy Committee, 1947–1977*. Ames: Iowa State University Press.
Szasz, Ferenc Morton (1984). *The Day the Sun Rose Twice*. Albuquerque: University of New Mexico.
Taton, René, ed. (1964). *History of Science: Science in the Twentieth Century*, translated by A. Pomerans. New York: Basic Books.

Teller, Edward (August 1991). Telephone interview with R. Howes.
Tennenbaum, Jonathan (1994). *Kernenergie—Die weibliche Technik.* Wiesbaden, Germany: Dr. Boettiger Verlags-GmbH.
Tinsley, Ellen L. (November 1991). Questionnaire.
Trbuhović-Gjurić, Desanka (1988). *Im Schatten Albert Einsteins: Das tragische Leben der Mileva Einstein-Marić.* Bern, Switzerland: Verlag Paul Haupt.
Troemel-Ploetz, Senta (1990). "Mileva Einstein-Marić: The Woman Who Did Einstein's Mathematics." *Women's Studies International Forum* 13, No. 5.
Truslow, Edith C. (1991). *Manhattan District History, Nonscientific Aspects of Los Alamos Project Y 1942–1946,* edited by Kasha V. Thayer. Los Alamos, NM: Los Alamos Historical Society.
Truslow, Edith C. and Ralph Carlisle Smith (1961). *Manhattan District History: Project Y: The Los Alamos Project,* Vol. II. Los Alamos, NM: Los Alamos Scientific Laboratory.
Ulam, S. M. (1976). *Adventures of a Mathematician.* New York: Charles Scribner's Sons.
University of Chicago Metallurgical Laboratory. "Report of Health Division for Month of May 1945," CH 2991. On the OHRE Web site (http://www.ohre.doe.gov).
U.S. Department of Energy/Martin Marietta Energy Systems Inc. (1992). "Wartime Laboratory." *Oak Ridge National Laboratory Review,* Nos. 3 and 4, chap 1, p. 25.
Van Arsdol, Ted (1958). "The City That Shook the World." Series of articles in *Columbia Basin News,* Richland, WA.
Van Assche, Pieter H. M. (1988). "The Ignored Discovery of Element $Z = 43$." *Nuclear Physics* A480, pp. 205–214.
Vare, Ethlie Ann and Greg Ptacek (1988). *Mothers of Invention.* New York: William Morrow.
Wakefield, Ernest (January 1994). Private communication with C. Herzenberg.
Walker, Evan Harris (February 1989). "Did Einstein Espouse His Spouse's Ideas?" *Physics Today* 42, pp. 9–11.
Walker, Evan Harris (February 1991). "Mileva Marić's Relativistic Role." *Physics Today* 44, pp. 122–124.
Walker, Evelyn S. (November 1991). Questionnaire.
Wallace, Dorothy A. (1962). "Further observations on Th-228 and Ra-226 in bones from dial painters and medical patients." Argonne National Laboratory publication ANL-EN98, pp. 63–66.
Walton, Anne (1989). "Lise Meitner and the Discovery of Atomic Fission" in *GASAT 5 Contributions to the Fifth International Conference 1989,* edited by Israela Ravina and Yael Rom. Haifa, Israel: Technion–Israel Institute of Technology.
Washburn, Patricia (1993). "Women and the Peace Movement" in *Women and the Use of Military Force,* edited by Ruth H. Howes and Michael R. Stevenson. Boulder, CO: Lynne Rienner Publishers.

Watkins, Sallie A. (1983). "Lise Meitner and the Beta-Ray Energy Controversy: An Historical Perspective." *American Journal of Physics* 51, pp. 550–553.

Watkins, Sallie A. (1983). "Lise Meitner's Scientific Legacy" in *Making Contributions: An Historical Overview of Women's Role in Physics*. College Park, MD: American Association of Physics Teachers.

Watkins, Sallie A. (1984). "Lise Meitner: The Making of a Physicist." *The Physics Teacher* 22, pp. 12–15.

Watkins, Sallie A. (1993). "Lise Meitner (1878–1968)" in *Women in Chemistry and Physics: A Biobibliographic Sourcebook*, edited by Louise S. Grinstein, Rose K. Rose, and Miriam H. Rafailovich. Westport, CT: Greenwood Press.

Watson, Annetta (1990). Private communication with C. Herzenberg.

Watson, Judy (September 29, 1995). Private communication with C. Herzenberg.

Way, Katherine (February 1991). Telephone interview with R. Howes.

Weaver, Ellen Cleminshaw (July 18, 1991). Telephone interview with R. Howes.

Weeks, Mary Elvira (1956). *Discovery of the Elements*, 6th Ed. Easton, PA: Journal of Chemical Education.

Weinberg, Alvin (February 5, 1992). Private communication with C. Herzenberg.

Whiting, Mary J. (November 1991). Questionnaire.

Williams, Laurie (October 31, 1993). "Women Came to Hanford to Make a Difference" and "Hanford's Mom." *Tri City Herald*, Kennewick, WA.

Wilson, Jane, ed. (1974). *All In Our Time*. Chicago: Bulletin of the Atomic Scientists.

Wilson, Jane S. and Charlotte Serber, eds. (1988). *Standing By and Making Do: Women of Wartime Los Alamos*. Los Alamos, NM: Los Alamos Historical Society.

Wise, Nancy Baker and Christy Wise (1994). *A Mouthful of Rivets: Women at Work in World War II*. San Francisco: Jossey-Bass.

Wood, Constance Simonsen (March 1996). Letter and telephone interview with R. Howes.

Yankovitch, Peter (July 28, 1995). Interview with R. Howes.

Yost, Edna (1960). *Women of Modern Science*. Cornwall, NY: Dodd, Mead.

Zirkle, Raymond E. et al. (1947). "The Plutonium Project," *Radiology* 49, No. 3.

# Index

Page numbers in italics indicate photographs.

Abelson, Philip H., 40, 70
academic positions, 59
accidents, 9, 53, 83
activism of scientists, 186–187
Adams, Gayle E., 138
administrative responsibilities, 152
African-Americans, 123, 135, 162, 163
Agnew, Harold, 190
Allison, Samuel K., 49, 189, 190
alpha particles, 22, 28, 90
aluminum alloys, 83, 84
Amacker, Hope Sloan, 165, 183, 184
American Chemical Society, 185, 192, 194
American Men of Science, 36
American Physical Society, 2, 35, 185, 191
Ames Project, 10
analytical chemistry, 73, 81, 133, 134, 137, 142–144, 146, 150
Anderson, Elda "Andy," 35, 60, *63*, 191–192
Anslow, Gladys Amelia, 18, 59
Anthony, Dave, 122
Apollo Project, 14
Argo, Harold, 3, 50
Argo, Mary Langs, 50–51, *63*
Argonne National Laboratory, 73, 77, 91, 102, 118, 119, 120, 124, 131, 132, 135, 138, 140, 191, 193, 194
Argonne Site, 41, 73, 123, 124, 194
Army, U. S., 11, 13, 17, 96, 123
Army Corps of Engineers, 9, 11, 173
Army Specialized Training Program, 132
artificially induced radioactivity, 20, 29
Association of Los Alamos Scientists, 187
Association of Manhattan Project Scientists, 187
Association for Women in Science, viii, 196
atomic bomb, 6, 16, 20, 22, 32, 35, 133, 137
Atomic Energy Commission, 76, 193, 195
atomic or fission piles. *See* nuclear reactors
Atomic Scientists of Chicago, 187

attitudes towards bomb, 5, 17, 43, 55, 181–183, 184–187, 199
attitudes towards women, vii, 11, 17–18, 24, 44–45, 81, 88, 94, 99–100, 108, 127, 134, 142, 146, 147, 169, 181–182, 199–200
attitudes towards work, 92, 103, 121, 133, 138–139, *175*, 182–183

Bachelder, Myrtle "Batch," 149–150
Bacher, Jean, 99
ban on publication, 7
Bardeen, John, 42
bargaining power, 16
barium, 6, 29, 31, 32, 86, 87
barium-140, 86
Barlett, Helen Blair, 67
barracks, 85, 141, 144, 162, 163, 165, 166.
  *See also* dormitories
Barschall, Heinz, 4
Baumbach, Harlan, 77
Baumbach, Nathalie Seifert, 77, *176*
Bayo Canyon, 87
Becquerel, Henri, 22, 23, 24, 25
Bederson, Benjamin, 56, 57
Bell, Iris, 165
Bench, Helen Landriani, 81
Berkeley. *See* University of California at Berkeley
Bernstein, Elaine Katz, 120
Berry, Yvette, 134–135, 142
beryllium, 28, 53, 84, 90, 129
beta decay, 57
beta radiation, vii, 21, 22, 25
Bethe, Hans, 20, 53, 105, 106
Bethe, Rose, 157
Bettelheim, Bruno, 105
binding energies, 30
biological effects of radiation, *66*, *67*, 91, 115, 122
biologists, 115–131
biomedical concerns, 115
biomedical technicians, 137, 183

253

births of babies, 41, 42, 49, 58, 154
Bisberg, Virginia Hawley, 73
bismuth phosphate carrier separation process, 79, 80
blacklisting, 188
Blacks. *See* African-Americans; segregation
blood tests, 119, 123, 124
blue badges, 172
Bohr, Niels, 7, 20, 28
bomb assembly, 84, 151
bomb construction, 9
bomb design, 9, 46, 84, 90, 97–98, 140; gun design, 46, 85, 97, 105; implosion, 46, 85–86, 97, 98, 105, 136; neutron reflector, 48–49
bomb testing, 93. *See also* Trinity
bootlegger, 166
boredom, vii, 15, 51, 142, 169
Born, Max, 37
Bower Award, 192
Bowman, Eleese, 140
Box 1663, 157
Boykin, Pearline (sometimes Perline), 135
Bradford, Rebecca, 146–147
B reactor, 128
Brewery, the, 73, 83, 118, 119. *See also* Site B
British atomic bomb project, 7, 12, 13, 33
British delegation to Los Alamos, 13, 32, 106, 197
Brode, Bernice, 99, 100, 103
Brode, Robert, 103
Brookhaven National Laboratory, 33
Brooks, Harriet, 21, 22
Brown, Evelyn J., 140
*Bulletin of the Atomic Scientists*, 173
Burkhardt, Bernice Morgan, 173
Burnett, Gladys S., 165
Bush, Vannevar, 8
Butcher, Mitizi Mars, 171
by-products, 73

cadmium, 41
calculating machines, vii, 93, 94, 98, 99, 101, 104, 132
calculators, 93–110
Calhoun, Opaline, 139
calutron operators, *111*, 169, 170
calutrons, 41, 78, *111–112*, 144, 168–169
Campbell, Miriam White, 150–151

Canadian Radium and Uranium Company, 90
Canadian scientists, 21
Carbide and Carbon Chemical Corporation, 95
careers in science, 137, 191, 200
Carnegie Institution, 8
Carritt, Jeanne, 147
Carrizozo, 49, 50
Carter, Robert, 55, *62*
Casler, Ruth A., 139
censorship, 17, 33, 167, 168
centrifugal separation of isotopes, 40, 41
cesium, 71, 72
Chadwick, Sir James, 28
chain reaction, 7, 8, 33, 37–38, 40, 43, 45, 46, 49, 54, 90, 97, 118. *See also* nuclear chain reaction
Chamié, Catharine, 21, 28
chelation therapy, 120
chemical engineers, 79, 80
chemical separation, 71
chemistry, 14, 18, 67–92, 132, 133, 134, 135
chemists, 67–92, 149
Cherenkov radiation, 54
Chicago Pile #1. *See* CP-1
child care, 14, 15, 42, 101, 126, 139, 154, 155, 162, 164, 196, 197
children, 15, 50, 99, 100, 101, 102, 122, 127, 162, 193
Chiotti, Wilma, 173
Christy, Robert, 53
Clark, Joan R., 95–96
classified information and documents, 9, 47, 89, 90, 145, 153, 161, 162
clearance. *See* security, clearances
clerical staff, 14, 17, 140, 152, 154, 171, 172
clinical programs, 125, 128
Clinton Engineer Works, 9, 10, 17, 58, *65*, 78–82, 95, *111, 112, 113,* 119, 124–128, 140, 144, 168, 169–170, *176, 177, 178, 179*
cloud chambers, 51, 52
Coatie, Georgia Mason, 162–163
Cockroft-Walton accelerator, 49
Cohen, Karl, 94
Coleman, Virginia Spivey, 81
Columbia River, 82
Columbia University, 8, 10, 12, 37, 38, 45, 47, 69, 72, 78, 88–89, 94, 95, 117, 191

Compton, Arthur Holly, 72, 125, 153, 183–184
computer development, 98, 109
computers (equipment), 93, 98. *See also* IBM electric calculating machines
computers (personnel), 96–110
Conant, James, 85
Condon, Edward, 187–188
construction projects, 140, 141
contamination. *See* radioactive contamination
control of atomic energy, 173, 181, 183, 186, 187, 189
Cook, Mary Jane, 127–128
Cooper, Cora F., 172, 196–197
core of bomb, 68
Cornell University, 8
corrosion, 12, 83
Cortelyou, Ethalaine, 76–77
Coryell, Charles, 58
Covey, Elwin, 140
Coveyou, Ida M., 171
CP-1 (Chicago Pile Number One), 10, 38–39, 41, 43, 119
Crawford, John A., 39
Crawford, Lorraine Golden. *See* Golden, Lorraine
critical assembly tests, 129, 130
criticality, 54
critical mass, 85, 97, 98
crystallography, 36
Curie, Eve, 25
Curie, Marie, 7, 21, 22, 23–26, 28, 33, 59, 61, 89
Curie, Pierre, 21, 23, 24, 25, 61
Curie Institute, 21
cyclotron group, 35
cyclotrons, 18, 41, 51, 52, 70, 73

Daghlian, Harry, 54–55
Daniels, Minnie M., 133, 138
Davies, Mrs. T. H., 128
Dayton, Jean Klein, 136–137
D-Day, 11
deferments, 12
DeGooyer, Marge Nordman, 3, 17–18, 142–143, 195
De Le Vin, Emma, 99, 102
demobilization, 181

Depression, Great, 59, 75, 104, 132, 152, 162, 181
designing atomic bomb, 46–47
desktop calculators, 94, 97–99, 100–102, 107
detonations, 97, 98
detonation systems, 136, 146
detonation waves, 86
detonators, 56
deuterium, 50
dieticians, 151
differential equations, 94, 102
disarmament movement, 5
discovery of radioactivity, 23
discrimination, vii, 58, 59–60, 75, 76, 81, 85, 94, 96, 104, 108, 127, 134, 146, 148, 152, 166, 181, 182, 189, 190, 191, 197, 198, 199, 200. *See also* segregation
dormitories, 43, 154, 170, *178*, 197. *See also* barracks
dose rates, 116–117
dosimetry, 116, 117, 128, 142
draftspersons, 150, 151
drivers, 152, 155, 164, 166
Duffield, Priscilla Greene, 156, 159–160, 161
Duffield, Robert, 160
Dumas, Louise, 165
Dunne, Frances, 19, 147–148, 195
Dunning, John, 10, 89
Du Pont Corporation, 13, 17, 42, 78–79, 83, 128, 141; medical programs, 128

Ehrlich, Eleanor Ewing. *See* Ewing, Eleanor
Ehrlich, Richard, *176*, 197
Einstein, Albert, 8, 12, 26–28, 187
Einstein, Mileva Marić. *See* Marić, Mileva
Elda E. Anderson Award, 192
electromagnetic separation of isotopes, 40, 78, 98, 168, 169. *See also* calutrons
electroscope, 71, 72
Elliott, Josephine, 99, 102
Emmerson, Harryette Hunter, 151
engineering, 14, 73
Engst, Erwin, 190
ENIAC (first digital computer), 108, 109
enriched uranium, 78
enrichment of uranium. *See* isotope separation
environmental monitoring, 128

Estabrook, Grace McCammon, 96, 145
European women scientists, 20–34
Evans, Cerda, 109
Evans, Foster, 109
Evans, Marjorie Woodard, 71
Ewing, Eleanor, 107, 108–109, *114*, *176*, 197–198
explosive lenses, 13, 85, 86
explosives, 14, 56, 85–87, 97, 148
Explosives Group, 19, 56, 148
explosives testing, 146, 148

faculty positions, 59
Failey, Crawford, 194
Failey, Hoylande Young. *See* Young, Hoylande
Failla, Gioacchino, 116, 117
Fat Man (plutonium implosion bomb), 148
Federal Bureau of Investigation (FBI), 89, 94, 134, 144, 147, 148, 195
Federation of American Scientists, 187, 188
feminist movement, 199, 201
Ferguson, Mrs. H. K., 18–19
Ferguson Engineering, 18–19
Fermi, Enrico, 8, 10, 12, 14, 20, 29, 30, 38, 39, 41, 45, 52, 53, 55, 60, *62*, 88, 118, 189–190, 192, 194; chain reaction, 8, 20; and Noddack, 29–30; young physicists, 55
Fermi, Laura, 158–159, 172, 185, 186
Feynman, Richard, 101, 106–107
film badges, 116, 117, 128
Fineman, Olga Giacchetti. *See* Giacchetti, Olga
Fineman, Philip, 133
Finkel, Asher Joseph, 118
Finkel, Miram Posner, 118, 193–194
Fisher, Leon, 185
Fisher, Phyllis, 185
fission, 6, 29, 32, 33, 35, 38 (*see also* nuclear fission); energy, 97; products, viii, 43, 53, 71, 72, 79, 80, 86, 118
Flanders, Donald "Moll," 99, 100, 101, 102, 103
Florin, Alan E., 133
Florin, Kay. *See* Gavin, Kathleen
Ford, Mary Rose, 127–128
Foreman, Beatrice, 139

Fornafelt, Elsie, 68
Fortenberg, Rosellen Bergman, 144
Fossey, Diane, 201
Foster, Margaret, 91–92
fractionation, 22
France, 12, 26
francium, 21
Franck, James, 183
Franck Report, 184
Frankel, Mary, 99, 100, 106
Frankel, Stanley, 98–99, 100, 105, 106, 108
Freedman, Mel, 4
Freeman, Elsie Mae, 133
Freeman, Julia Elaine, 140
French, Naomi Livesay. *See* Livesay, Naomi
French, Tony, 197–198
French atomic bomb project, 7
Friedlander, Gerhardt, 87
Frisch, Otto, 6, 7, 30, 32, 55, 136
Frisch, Rose Epstein, 131
Fuchs, Klaus, 13, 95, 197, 198
Fuchs, Mrs. Robert. *See* Melhase, Margaret
fuel rods, 71
Fuian, Emil, *64*
Fuller Lodge, 109, 136, 197
fusion weapons, 47, 50, 51

Gailar, Joanne Stern, 171
Galuskin, Sylvia, 164
Garvey, Kay, *64*
gaseous diffusion, 19, 37, 40, 45, 68–69, 70, 72, 78, 79, 144
Gaston, Evelyn, 119
Gavin, Kathleen, 133, 138, *176*
Geiger counters, vii, 44, 49, 50, 54, 58, 130, 136, 151
German atomic bomb project, 7, 8, 95
German physicists, 33
Germany, 17, 31
Giacchetti, Olga, 75, 132–133
Giesel, Frederick, 22
Gilbreath, Rachel, 139
Gish, Eleanor, 39
glass ceiling, 18. *See also* discrimination
Göttingen, 20
Golden, Lorraine, 39–40
Goldhaber, Gertrude Scharff, 33
Goldowski, Nathalie Michel, 12, 83–84, 189

Gordon, Pearl Leach, 130
graduate students, 41, 48, 50, 51, 55
graphite, 8, 38
Graphite Reactor, viii, 80
Graves, Al, 48, 49, 50, 130, 193
Graves, Elizabeth "Diz" Riddle, 48–50, 63, 130, 193
Greene, Priscilla. See Duffield, Priscilla Greene
Greisen, Kenneth, 146
Grieff, Lotti, 69
Gross, Norma, 86–88, 149, 166–167, 183, 192
Groves, General Leslie, 9, 17, 41, 47, 65, 78, 93, 101, 132, 146, 152, 154, 160, 162, 163, 168
Guadagna, Lillian, 139–140
gun design. See bomb design

Hahn, Otto, 6, 21, 30, 88; and Meitner, 31–32, 62
Hall, David, 43
Hall, Helen, 170
Hall, Jane Hamilton. See Hamilton, Jane
Hamilton, Jane, 43–45, 129, 187, 194–195
Hammer, Geneva Owen, 163, 171
Hammermesh, Mort, 102
Hanford Engineer Works, 9, 10, 13, 17, 18, 43, 44, 45, 75, 81–84, 119, 128, 129, 137, 140–144, 154, 162–163, 178, 179, 183, 184
Hanig, Evelyn Kalichman, 81
Happer, Gladys Morgan, 125–127
Happer, William, 3, 125, 126
Harding, Reid, 64
Harvard University, 25
Harvey, Kay, 81, 140
Hauk, Eleanor, 65, 144–145
Health Division, 118, 120
health effects of radiation, 73, 118
health physics, 44, 129
Health Physics Society, 192
health risks, 137
Heimbach, Dave, 168
hematology, 131
Hemmindinger, Peggy, 15
hemolytic effects of radiation, 120, 121
Hempelmann, Louis, 172
Herrick, Susan Chandler, 69–70
Heydorn, Jane, 165

Hillhouse, Dorothy, 173
Hinton, Joan, 3, 11, 51–56, 62, 155, 187, 189–190
hiring. See recruiting women
Hiroshima, 1, 5, 11, 43, 46, 55, 148, 151, 181, 184, 185, 186, 189, 199; attitudes toward, 130, 170
Hirschfelder, Joseph, 103, 104, 105
Histories of the Manhattan Project, 2, 173
Hitler, 6, 20
Hoekstra, Henry, 140
Hoekstra, Marilyn. See Jordan, Marilyn "Jodie"
Holmes, Virginia S., 72
home life on the Manhattan Project. See living conditions
honorary doctorates, 25
Hornig, Donald, 85
Hornig, Lilli, 84–86, 185
hospitals, 66
hot labs, 86
housing shortages, 16
Howe, John, 78
Howe, Marilyn, 78
Howes, Bob, 4
Howes, Jane, 4
Hughes, Kathleen, 76
Hume, David, viii
humor, 44, 49, 91, 103, 134, 139, 153, 154
hydrodynamics, 98
hydrogen bombs. See fusion weapons
Hypo Reactor, 53

IBM electric calculating machines, 98, 99, 100, 104, 105, 106, 107, 109, 132
implosion, 86; calculations, 106, 107, 108
implosion bomb, 98; designs (see bomb design)
initiators, 46, 90
Inglis, Betty, 99, 102, 173
Inglis, Dave, 102
instrumentation, construction of, vii–viii, 24, 51
interface calculations, 106
interviews. See recruiting women
ionization chambers, 22, 24, 28
Iowa State College, 8, 10, 68, 173
isotope enrichment. See isotope separation

isotope separation, 7, 9, 10, 40, 41, 68, 69, 78, 84, 98, 140, 144

Jaffe, Art, 123
Japan, 181
Japanese surrender, 107
Jennings, Nellie, 140
Jensen, E. Ann, 173
Jette, Eleanor, 15, 16, 154–155
Jette, Eric, 155
Johns Hopkins University, 8, 37
Johnson, Margaret, 99, 102
Joliot-Curie, Frédéric, 7, 22, 28–29, 33, *61*
Joliot-Curie, Irène, 7, 20, 21, 22, 28-29, 33, 34, 60, *61*, 88
Jones, Roberta Harvey, 130
Jordan, Marilyn "Jodie," 140
Jumbo (steel vessel designed to contain explosion and conserve plutonium), *114*
Jupnik, Helen, 57–58

K-25 facility (gaseous diffusion plant at Clinton), 79, 145, 171
Kaplan, Irving, 94
Karle, Isabella Lugoski, 74–75, 192
Karle, Jerome, 74–75, 192
Katz, Joseph, 171
Keck, James, 57
Keck, Margaret Ramsey. *See* Ramsey, Margaret
Kellex (a susidiary of M.W. Kellogg Company), 10
Kennedy, Joseph, 70, 135
Kenny, Ann, 96
Kerst, Donald, 53
Kirkley, Ada, 144
Kirtland Air Force Base, 148
Kistiakowsky, George, 20, 85, 87, 148
Knickerbocker Stables at B site, 119, 120
Koshland, Daniel, 80
Koshland, Marian Elliott, 80
Koskosky, Adele, 139
Koziolek, Winifred T., 139
Kurath, Dieter, 102, *176*
Kurath, Frances Wilson. *See* Wilson, Frances

labor force, 13, 14, 141, 146, 149, 154
labor shortages, 11, 14, 18, 52, 132, 141, 148, 155, 164

labor unions, 13
Lamy, 108, 167
Lane, Mary, 172
Langer, Beatrice, 99, 100, 102
Langer, Rudolph, 104
Langevin, Paul, 26
Langs, Mary. *See* Argo, Mary Langs
lanthanum-140, 86–87
lanthanum fluoride process 79, 82
Lathrop, Katherine L., 121–123, 193
Latimer, Wendell, 72
Lawrence, Ernest, 41, 45, 159, 168
Lazarus, Marcella D., 173
lead-210, 89
lead bricks, vii
Lefkowitz, Hilda, 94–95, 196
Lemke, Valda B., 140
Lenoff, Gladys, 128
Lewis, Eleanor, 139
Lewitz, Elaine, 140
Leyshon, Emily, 81
Libby, Leona Woods Marshall. *See* Marshall, Leona Woods
Libby, Willard, 193
librarians, 15, 159, 160, 173
Little Boy (uranium gun-type bomb), 148, 151
Livesay, Naomi, 2, *63,* 102, 104–106, 107, 108, 109, 197
living conditions, 14–17, 85, 103, 105, 126, 129, 135, 136, 139, 140, 141, 144, 146, 155, 157, 158, 159, 160, 162, 163, 166, 167, 170, 182
Long, Mrs., 173
Lopo Reactor, 53
Los Alamos, 9–10, 11, 13, 14–15, 16–17, 20, 43, 45, 47–48, 50–52, 55, 58, *65,* 84–88, 99–109, 129, 135–136, 140, 146–151, 152–162, *175, 176,* 182, 184, 185, 189, activism, 186–187; baby boom, 154
Los Alamos Association of Scientists, 186

Mademoiselle Merit Awards, 1946, *180*
Maeder, Elmo, 165
Maeder, Irma Dowd, 165
maid service, 14, 99, 101, 136, 155
mail, 159, 160
Manhattan Engineer District, 9, 11

Manhattan Project, 1, 5, 6, 8, 9, 11, 14, 19, 20, 78; costs, 11
MANIAC computer, 109
Manley, John, 101, 161
Manley, Kay, 99, 101, 187
Maoist revolution, 190
Marić, Mileva, 26–28, 34
Maris, Buena, 17, 156, 162–164, *179*
Marks, Edna K., 119
marriage, 41, 134, 137, 158, 160, 166, 171, 176, 197; and women scientists, 36, 38, 39, 40, 48, 69, 70, 72, 127, 133, 134, 136, 139, 194, 196
Marshak, Robert, 56
Marshall, John, 38, 41, 42
Marshall, Leona Woods, 1, 38, 39, 41–42, 63, 166, *180*, 183, 192–193, 198
Massachusetts Institute of Technology. *See* MIT
mass-energy equivalence, 28
mass spectrometers, 10, 142, 144, 168
mathematicians, 93–110
Matthias, Colonel Franklin "Fritz," 163
Matthias, Mrs. Franklin, 184
Maxwell, Dr. Elizabeth, 147
Mayer, Joseph, 37
Mayer, Maria Goeppert, 12, 36–37, 40, 47, 50, 59, *62*, 69, 70, 102, 187, 191
McAndrew, Pat, 167–168, 195
McCamey, Creola Green, 169–170
McCarthy era, 188, 189, 199
McChesney, Marilyn, 147
McKibben, Dorothy, 156–160, 172, 182, 187
McMillan, Edwin M., 70
medical staff, 115–131
medical technicians, 147
Meitner, Lise, 1, 6, 7, 20, 21, 22, 29, 30–34, 59, 60, *62*, 88, 198; attitude towards bomb, 32; press coverage, 1
Melhause, Margaret, 71–72
Memorial Hospital, 116, 117
Meschke, Virginia, 78, 139
metallography, 136
Metallurgical Laboratory, 10, 36, 38, 39, *64*, 72–78, 81, 82, 83, 117–124, 135, 137–140, 153, *176*, *178*, 181, 183–184, 191
metallurgy, 12, 14, 73, 83, 147
Met Lab. *See* Metallurgical Laboratory
Metropolis, Nicholas, 106, 107, 109

military discipline, 148, 166, 167
military intelligence, 165
military involvement, 6, 7, 9
military personnel, 13. *See also* SED, WACs
Miller, Mary Lucy, 149
Millikan, Robert, 59
Mindlin, Raymond, 95, 196
MIT, 9, 34, 67, 70, 173
Mitchell, Dana P., 156–157, 174
mobility restrictions, 108, 146
Mokstad, Betty, 139
Monet, Marion Crenshall, 80
Monk, Ardis, 96
Monsanto Chemical Company, 81
monte-carlo calculations, 109
Mooney-Slater, Rose Camille LeDieu, 36
Morgan, Karl Ziegler, 125, 126, 127
Morrison, Philip, 53
motherhood, 15, 155. *See also* birth of babies, childcare, children, wives
movement to control atomic energy, 183–190
Mrozowski, Stanislaw and Irena, 12
Muncy, J.A.D., 157
Murray, Betty, 133
Museum of Science and Industry, 73
music, 55, 155
Myers, Dorothy, 174

Nachtrieb, Mary, 85
Nachtrieb, Norman, 150
Nagasaki, 5, 46, 53, 55, 148, 181, 185, 186, 199
National Academy of Sciences, 25, 33
National Bureau of Standards, 8
National Defense Research Committee. *See* NDRC
National Defense Research Council. *See* NDRC
national laboratories. *See specific names of laboratories*
National Nuclear Energy Series, 76
National Research Council, 57
Native Americans, 124, 130, 136, 194
natural radioactivity, 25
*Nature*, 32
Naval Research Laboratory, 40, 192
Navy, U. S., 8

NDRC (National Defense Research Committee; later the National Defense Research Council), 8, 18
Nelson, Eldred, 98–99, 105
nepotism rules, 48, 192. *See also* discrimination
neptunium, 41, 42, 70, 138
Nereson, Jean Parks, 173
neutron: absorption, 38; capture, 32; cross-sections, 43; detectors, 38; multiplication experiments, 38; reflectors, 48; scattering, 49
neutrons, 23, 28, 29, 30, 31, 33, 90, 98, 150; fast, 49; slow, 38; thermal, 68
Newman, Mary Holiat, 69
New Mexico, 1, 9, 11, 51, 85, 108, 157, 159, 168
Nichols, Colonel Kenneth, 153
nickel barriers, 70, 79
Nickson, James J., 119
Nickson, Margaret J., 119
Nier, Albert, 10
niobium, 139
Noah, Frances E., 102
Nobel Prize, 20, 25, 28, 30, 32, 42, 47, 74, 94, 101, 191
noble gases, 45
Noddack, Ida Tacke, 21, 29–30, 34
Noddack, Walter, 21, 30
Nordheim, Gertrud, 58–59
Nordheim, Lothar, 58
Novey, Elaine. *See* Lewitz, Elaine
Novey, Ted, 140
nuclear chain reaction, 1, 10, 37, 38
nuclear explosion, 37
nuclear fission, 37,; discovery of, 7, 23, 29–32; early experiments on, 33; energy release, 7, 26, 28
nuclear isomerism, 31
nuclear physics, 20, 73, 140
nuclear power, 5
nuclear reactors, 9, 33, 37, 38, 39, 41, 45, 72, 82, *114*, 140; construction of, 42, 43, 128, 141; enriched uranium, 52–53
nuclear science, 20, 21
numbers of women, 13–14
numerical simulations, 93, 98, 101, 106, 107
nursery schools. *See* child care

nurses, *66*, 82, 128, 129, 131, 142
Nyden, Shirley, 139

Oak Ridge, vii, 4, 9, 19, 42, 43, 58, 69, 77, 78, 79, 80, 95, 117, 124, 126, 127, 137, 140, 144–146, 168–170, *176*, *177*, *178*, *179*. *See also* Clinton Engineer Works
Oak Ridge Hospital, *66*
Oak Ridge National Laboratory, 145, 191
O'Bryen, Brian, 89–91
Office of Human Radiation Experiments of the Department of Energy, 91
Office of Scientific Research and Development, 18, 59, 80, 90, 93, 121
OHRE. *See* Office of Human Radiation Experiments of the Department of Energy
old-girl networks, 3
O'Leary, Jean, *65*, 152–153
Olsson, Virginia, 153
Omega site, 53, 54
opacity calculations, 47, 50, 102
Oppenheimer, J. Robert, 2, 9, 47, 101, 150, 156, 158, 159, 160, 161–162, 184, 188
Oppenheimer-Phillips reaction, 188
optics research, 48
ordnance, 14
Oregon State College, 17
OSRD. See Office of Scientific Reserch and Development

packing fraction, 32
Painter, Elizabeth, 121
Palevsky, Harry, 48
Palevsky, Elaine Sammel. *See* Sammel, Elaine
Paris, 33
Parker, Herbert, 44
Parsons, Aquilla, 133
Parsons, Captain William S., 158
Pasco, Washington, 17
patriotism, 15, 17
Patterson, Pat, 170
pay, 15, 133, 134, 141, 156, 162, 169, 173, 182, 197. *See also* salaries
Pearl Harbor, 101
Peierls, Rudolph, 95
Pellock, Helen, 134
Perey, Marguerite, 21

Perley, Anne, 129, 130
Perrin, Jean Baptiste, 24
personnel shortages. *See* labor shortages
Peterson, Sigfred, 134
Philadelphia Navy Yard, 19, 40
Phillips, Anne, 153
Phillips, Melba, 2, 187–188
physicians, 125, 126, 128
*Physical Review*, 150
physicists, 14, 35–60
photochemical reactions, 40
photographic emulsions, 150
photomicrographs, 56
physiological effects of ionizing radiation, 121
Pierce, Elsie, 164, 165
Pinckard, Marian, 139
pipetting, 135, 143
pitchblende, 24
plutonium, 41, 53, 54, 70, 71, 72, 73–74, 75, 84, 93, 118, 120, 128, 129, 137, 138; chemistry, 74, 85; compounds, 74, 77, 79; inhalation of, 44; production of, 9, 42, 71, 78–80, 81–83, 142, 143, 144, 146, 147; research, 72; separation of, 79, 81, 138
plutonium-239, 97, 129, 138
plutonium-240, 46, 53, 54, 98
plutonium bomb, 46, 55, 84, 85, 86, 90, 97
plutonium project, 76
poisoning of reactors, 43, 45
polonium, 12, 22, 24, 88–91, 129
Pomerance, Eleanor Eastin Hauk. *See* Hauk, Eleanor
Pomerance, Herbert, 3
Porter, Lillie Mae, 120–121, 133
positrons, 22, 28
predetonation of weapon, 33, 84, 98
pregnancies, 41, 49. *See also* births of babies
Priest, Grace, 81
Princeton University, 25, 45, 57, 96
production reactors, 9, 12, 40, 41, 42, 43, 44, 45, 78, 79, 81, 82, 84, 128
profanity, 42
Project Alberta, 19
promotions, 15, 151, 190
protactinium, 21

quartz-fiber balances, 147
Quimby, Edith Hinkley, 22, 115–117, 131, 190–191
Quimby, Shirley L., 116

racetracks, 144, 168
radar, 11, 34, 35
radiation: accidents, 54, 55, 129, 130, 139, 193; alarms, 54; chemistry 73, 79; dosimetry, 22; exposure, 54, 87, 128, 129, 172; handling, 65, 79, *113*; hazard, 55, *113*, 128, 129, 133, 137–138, 142, 143; monitoring, 49, *113*, 116, 117, 120, 128, 129, 133, 142; protection, 44; shielding, vii; sickness, 55; symbol, 145
radioactive clouds, 50
radioactive contamination, 79, 87, 123, 126, 135, 142, 143
radioactive decay, 21, 24
radioactive emissions, 128
radioactive spills, 39, 40
radioactive waste, 79, 128
radioactivity, 22, 23, 24, 31
radioassays, 146, 147
radiochemical analyses, 141
radiolanthanum implosion experiments. *See* lanthanum-140
radium, 21, 24, 25, 31
radon, 28, 89
Ramsey, Margaret, 56–57
Ray, William, 65
Reace, Eleanor, 70
reaction to Hiroshima, 184–185, 186
reactor accidents, 53
reactor constants, 43
reactor design, 42, 82
reactor safety, 44
reactors. *See* nuclear reactors
Read, G. M., 163
recruiting women, 3, 10–14, 15, 132, 133, 134, 136, 141–142, 146, 152, 153, 162, 169, 173
Rees, Mina, viii, 18, 93
refugees, 12
relativity theory, 26
research, 43, 44
resonance absorption of neutrons, 58
rhenium, 21, 30, 58
Richland, Washington, 9

Ride, Sally, 201
Ringo, Roy, 4
Rizzoli, Ida, 173–174
Roberg, Jane, 57, *63*
Robinson, Donna, 143–144
Robson, Arthur, 183
Robson, Melba, 119, 124–125, 183
Roentgen, Wilhelm, 23
Rona, Elizabeth, 12, 88–91
Roosevelt, Franklin D., 8
Rosenthal, Marcia White, 120
Rosie the Riveter, 17, 181
runaway reactions, 41, 129, 130
Russell, Bertrand, 42
Russell, Mrs., 128
Rutherford, Ernest, 20, 21, 28

Sachs, Robert, 8, 191
safety, 54, 81–82
salaries, 15, 106, 147, 152, 162, 191, 197. *See also* pay
Sammel, Elaine, 48
Santa Clara indian pueblo, 130
Santa Fe, 16, 49, 102, 108, 130, 149, 156, 157, 158, 159, 160, 161
Sarah Lawrence College, 37
Saunders, Addie, 140
Schiedenhein, Captain Arlene, 166, *180*
Schmidt, Gerhard, 23
schools, 14, 154
Schulkin, Florence Pachter, 164, 196
Schwartz, Bernice, 94
science careers, 136, 143
scientific assistants, 102
scientists' organizations, 186–187
Seaborg, Glenn, 4, 70, 71, 74, 77, 123, 134, 137, 138
Seaborg, Helen, 76
Seattle, 17
secrecy, vii, 73, 95, 115, 118, 124, 126, 133, 134, 137, 144, 145, 147, 148, 149, 151, 152, 158, 160, 161, 171, 181, 182, 184
secretaries, 14, 140, 152, 155, 159, 164, 171, 172
security, vii, 9, 12, 13, 15, 16–17, 33, 47, 55, 58, 72, 73, 87–88, 89–90, 94, 95, 97, 99, 108, 109, 118, 123, 131, 134, 137, 141, 143, 145, 147, 148, 151, 156, 157, 158, 159, 160, 161–162, 165, 167–168, 171, 195, 197; clearances, 12, 33, 47, 152, 153, 170, 172, 189; voluntary ban on publishing, 33
SED (Special Engineer Detachment), 13, 56, 99, 100, 106, 132, 133, 148, 164, 173, 182
segregation, 141, *177*
separation plants, 9, 82, 128, 140, 141, 142; construction of, 141. *See also* plutonium, separation of
Serber, Charlotte, 15–16, 155, 156, 159, 160–162, 182
Serber, Robert, 160, 161
Shanklin, Mary M., 165
shaped explosive charges, 46
Shaw, Margaret, M., 17, 129
Sheviakov, George, 105
shock waves, 97, 98, 106
Shopp, Marjorie Powell, 164, 195
Shor, Roberta, 81
Shupp, Selma, 139
Simonsen, Constance, 135–136
Site A, 73. *See also* Argonne Site
Site B, 118, 119, 120, 124. *See also* Brewery, the
Sites, Anna Mary, 174
Site X. *See* Clinton Engineer Works; Oak Ridge
Site Y, 159, 162. *See also* Los Alamos
Skyrme, Tony, 106
Slater, John Clarke, 36
Slater, Rose. *See* Mooney-Slater, Rose Camille Le Dieu
Slotin, Louis, 129–130, 193
slow neutrons, 38
slug canning, 84
Smith, Alice Kimball, 172–173
Smith, Barbara J., 145–146
Smith, Cyril Stanley, 173
Smith, Ralph, 173
Smith College, 18, 45, 59
Sohn, Madeline, 77
South Mesa, 56
Soviet atomic bomb project, 7, 95, 201
Special Engineer Detachment. *See* SED
special studies, 44
Speck, Lyda, 150
spectroscopy, 150
Spedding, Frank, 10, 68

Spence, Rod, 87
spontaneous nuclear fission, 33, 46, 90, 98
Stagg Field, 38, 73
Stahl, Margaret Smith, 185
Standard Oil Development Company, 8
standards of conduct, 154
Stark, Helen, 81
Starner, Bill, 54
statisticians, 145, 160
Steele, Frances E., 165
Stein, Amanda Pauline Blume, 77
Stimson, Henry, 183–184
Stockholm, 32, 33
Stone, Jerry, 166
Strassmann, Fritz, 6, 30, 31–32
strontium, 118
Stroud, Agnes Naranjo, 130–131, 194
Summers, Mildred M., 123–124
supercritical mass, 46
supervisors, 82, 101
Sweden, 6, 31
switchboard operators, 152, 165
Szilard, Leo, 8, 12, 38, 83, 88, 184

table of isotopes, 76
Tacke, Ida. *See* Noddack, Ida Tacke
Tasseneau, Janet, 120
teachers, 14, 15, 155, 172, 173
technetium, 21
technicians, 14, 18, 78, 132–151
technology, 44
telephones, 159, 160
Teller, Edward, 4, 8, 12, 37, 47, 50, 55, 109
Teller, Augusta "Mici," 99, 102, 109
Tennessee, 9, 80, 124, 169
Tennessee Eastman Corporation, 77, 78, 144, 145, 168, 169
Tenney, Edith, 131
Theoretical Division, 98, 99, 102, 105
Theoretical Group, 20, 50, 100
thermal diffusion, 19, 40–41, 78
thermonuclear fusion. *See* fusion
Thomson, Helen, 138
thorium, 23, 24, 92, 138
Tinian, 1, 19
Tinsley, Ellen L., 172
Tolmach, Emma, 139
Towle, Virginia, 139
toxicology of radionuclides, 118, 121

Tracy, Kay 153
training, viii, 135, 136, 141, 142, 143, 146, 147, 169
transuranic elements, 30, 39, 74, 118, 137
Trinity test, 1, 46, 49–50, 51, 54, 55–56, 87, 106, 130, 150, 151, 158, 184, 187, 190; reaction to, 124, 183
tritium, 50
truck drivers. *See* drivers
Trujillo, Isabelle, 174
Truslow, Edith C., 173
tubealloy, 73

Ulam, Stanislaw, 52
Union Carbide, 79
University of California, Berkeley, 8, 10, 41, 70–72, 73, 98, 144, 168
University of California, Los Angeles, 176
University of Chicago, 8, 10, 12, 36, 38, 39, 78, 102, 105, 118, 123, 150, 189, 192. *See also* Metallurgical Laboratory
University of Illinois, 33
University of Minnesota, 8, 10
University of Rochester, 89, 90, 91
University of Tennessee, viii
University of Virginia, 8, 41
University of Wisconsin, 11, 35, 51, 52
uranium, 6, 7, 8, 31, 32, 58, 68, 84, 136, 137, 147, 168; alloys, 70; chemistry, 70; compounds, 23, 24, 70; enriched, 84; enrichment, 45 (*see also* isotope separation); hexafluoride, 10, 19, 40, 69
uranium-233, 138, 139
uranium-235, 7, 10, 19, 35, 38, 40, 41, 46, 68, 69, 78, 84, 93, 97, 129, 168; abundance, 7; fission in, 7
uranium-238, 19, 40, 41, 68, 70, 168
Uranium Advisory Committee, 8
uranium project, 12
Urey, Harold, 37, 45, 68, 69

Van de Graaff accelerators, 52, 150
von Neumann, John, 107, 108, 109, 136, 151
von Neumann, Klara "Klari," 109

WAACs. *See* WACs
WACs (Women's Army Corps), 1, 4, 13, 82, 86, 132, 141, 148–151, 156, 157, 164–168, *175*, *178*, 182

wages. *See* pay; salary
Wagner, Juanita "Billie," 81
Wagner, Roberta, 77
Wahl, Joseph, 70
Walker, Evelyn S., 149
Wallace, Dorothy, 120
Walsh, Patricia, 39
Warshaw, Sylvia (sometimes Silvia), 77–78
Water Boiler reactor, 52, 55
Watts, Ellen, 139
Way, Katharine "Kay," 42–43, 187, 189, 191
Way-Wigner formula, 43
weapons design, 72. *See also* bomb design
weapons testing, 93, 136
Weaver, Ellen Cleminshaw, vii–viii, 80–81, 195–196
Weaver, Harry Edward, Jr., 80
weighing of plutonium, 73
Weinberg, Alvin, 42
Weiner, Sonia, 171
West Stands, 38, 102, 123
Wheeler, John, 42
White, Sergeant Miriam, 166
white badges 85, 87, 109
Whiting, Mary J., 164
Wigner, Eugene, 8, 12, 171
Wilder, Edward, Jr., 13
Wilson, Elizabeth, 149
Wilson, Frances, 99, 102, *176*
Wilson, Jane, 173
wives, 14, 15, 99, 100, 101, 102, 103, 136, 140, 153–154, 159, 169, 182
women, vii, viii, 1, 2
women managers, 82, 129, 146, 148, 149, 150, 151, 153, 160, 166, 195
Women's Army Corps. *See* WACs
womens' colleges, 59.

womens' movement. *See* feminist movement
Wood, David, 136
Wood, Nancy, 39
Woods, Leona. *See* Marshall, Leona Woods
working conditions, 15, 17, 120, 134, 136, 138, 141, 142, 143, 144, 145, 146, 151, 163, 170, 171, 173, 174, 182, 183
working women, vii, viii, 14–15, 154–155
work schedules, 99
work week, 16, 141, 181
World War II, 11, 20, 21, 52; end, 184, 185
WPA mathematical tables, 102
Wright, Edith, 99, 102
Wu, Chien-Shiung, 45–46, *64*, 191

X-10 facility (reactor at Clinton), 79, 80, 112, *113*
xenon, 45
xenon effect. *See* poisoning of reactors
X-ray crystallography. *See* crystallography
X-ray photography, 86
X-rays, 23

Y-12 facility (electromagnetic isotope separation plant), 78, 81, *111*, *112*, 144, 145, 170, *179*
Yale University, 25
Yalow, Rosalind Sussman, 94
Yang, C. N., 190
Yankovitch, Peter, 18, 72
Young, Bob, *64*
Young, Hoylande Denune, 75–76, 187, 194
Yuan, Luke Chia Liu, 45

Zachariasen, W. H., 35, 36